Intelligent Systems Reference Library

Volume 110

Series editors

Janusz Kacprzyk, Polish Academy of Sciences, Warsaw, Poland
e-mail: kacprzyk@ibspan.waw.pl

Lakhmi C. Jain, University of Canberra, Canberra, Australia;
Bournemouth University, UK;
KES International, UK
e-mails: jainlc2002@yahoo.co.uk; Lakhmi.Jain@canberra.edu.au
URL: http://www.kesinternational.org/organisation.php

About this Series

The aim of this series is to publish a Reference Library, including novel advances and developments in all aspects of Intelligent Systems in an easily accessible and well structured form. The series includes reference works, handbooks, compendia, textbooks, well-structured monographs, dictionaries, and encyclopedias. It contains well integrated knowledge and current information in the field of Intelligent Systems. The series covers the theory, applications, and design methods of Intelligent Systems. Virtually all disciplines such as engineering, computer science, avionics, business, e-commerce, environment, healthcare, physics and life science are included.

More information about this series at http://www.springer.com/series/8578

Seiki Akama

Editor

Towards Paraconsistent Engineering

 Springer

Editor
Seiki Akama
Kawasaki
Japan

ISSN 1868-4394   ISSN 1868-4408   (electronic)
Intelligent Systems Reference Library
ISBN 978-3-319-82095-8   ISBN 978-3-319-40418-9   (eBook)
DOI 10.1007/978-3-319-40418-9

Printed on acid-free paper

This Springer imprint is published by Springer Nature
The registered company is Springer International Publishing AG Switzerland

Foreword

In classical and in most non-classical logics, if a theory T is inconsistent, i.e., contains contradictory theorems, then it is also trivial; T is said to be trivial if any sentence of its language is provable in T. A theory founded on a paraconsistent logic may be inconsistent but non-trivial. So, we are able, with the help of such a logic, to develop inconsistent but non-trivial theories.

Paraconsistent logics can be conceived as logics that are rival of classical logic or as formal tools that may complement classical logic in certain situations.

One relevant point is that paraconsistent logics did find an extraordinary number of applications, which constitute the basis of applied paraconsistency. Leaving aside philosophy, law, and the foundations of science, paraconsistent logic came to be significant in domains like the following:

- Computer science:

  Artificial Intelligence (common sense reasoning, knowledge representation)
  Database, knowledge bases
  Data mining
  Conceptual analysis
  Software engineering

- Engineering:

  Signal processing (sound processing, image processing)
  Neural computing
  Intelligent control
  Robotics
  Traffic control in large cities

- Economics:

  Decision theory
  Game theory
  Finances

- Linguistics:

  Formal semantics
  Computational linguistics

Jair Minoro Abe was one of my best Ph.D. students at the University of São Paulo in the eighties of the last century. At that time, we did not know a large number of real and good applications of paraconsistent logic (envisaged as a rival of or as a complement to classical logic). Since the beginning of his Ph.D. work, Abe became interested in the possible applications of paraconsistent logic. He was, above all, one of the pioneers of this domain, little by little opening new ways, particularly in paraconsistent robotics, decision theory, and neural nets. After his contact with the Japanese logicians Seiki Akama and Kazumi Nakamatsu, the progress in the area of applications became wide and profound. In all applications involving a large part of technology, Abe, Akama, Nakamatsu, and collaborators employed paraconsistent logic as an instrument to cope with special and significant problems.

This book, dedicated to Jair Minoro Abe on the occasion of his 60th birthday, shows, at least in outline, a small part of what has been done in the field of applications of paraconsistency. On the other hand, it also makes clear how relevant and productive may be the collaboration between the experts of two distant countries, in this case Brazil and Japan.

Most papers collected in this book are related to applied paraconsistency. Moreover, all of them are dedicated to Jair Minoro Abe, the friend, the man, and the logician.

Curitiba, Brazil   Newton C.A. da Costa
January 2016

Preface

Paraconsistent logics refer to non-classical logical systems, which can properly handle contradictions. They can overcome defects of classical logic. Initially, they were motivated by philosophical and mathematical studies, but they recently received interesting applications to various areas including engineering. To tolerate contradictions is an important problem in information systems, and paraconsistent logics can provide suitable answers to it. Paraconsistent engineering, which is engineering based on paraconsistent logics, should be developed to improve current approaches. Jair Minoro Abe is one of the experts on Paraconsistent Engineering, who developed the so-called annotated logics. This book collects papers by leading researchers, which discuss various aspects of paraconsistent logics and related logics. It includes important contributions on foundations and applications of paraconsistent logics in connection with engineering, mathematical logic, philosophical logic, computer science, physics, economics, and biology. It will be of interest to students and researchers, who are working on engineering and logic. The structure of this book is as follows.

Chapter 1 by S. Akama gives an introduction to this book.

Chapter 2 by S. Akama and N.C.A. da Costa discusses the reason why paraconsistent logics are very useful to engineering. In fact, the use of paraconsistent logics is a starting point of Abe's work. The ideas and history of paraconsistent logics are reviewed. The chapter is also useful for readers to read papers in this book.

Chapter 3 by N.C.A. da Costa and D. Krause is concerned with an application of a paraconsistent logic to quantum physics. The paper reviews the authors' previous papers on the concept of complementarity introduced by Bohr. Logical foundations for quantum mechanics have been worked out so far. The paper reveals that the authors' paraconsistent logic can serve as the basis for the important problem in quantum mechanics.

Chapter 4 by J.-Y. Beziau proposes two three-valued paraconsistent logics, which are 'genuine' in the sense that they obey neither $p, \neg p \vdash q$ nor $\vdash \neg(p \wedge \neg p)$. Beziau investigates their properties and relations to other paraconsistent logics. His work is seen as a new approach to three-valued paraconsistent logics.

Chapter 5 by S. Akama surveys *annotated logics* which have been developed as paraconsistent and paracomplete logics by Abe and others. The paper presents the formal and practical aspects of annotated logics and suggests their further applications for paraconsistent engineering.

Chapter 6 by J.I. da Silva Filho et al. discusses an application of the annotated logic called *PAL2v* based on two truth-values for *paraconsistent artificial neural network* (PANet), showing an algorithmic structure for handling actual problems. The paper is one of the interesting engineering applications of paraconsistent logic.

Chapter 7 by K. Nakamatsu and S. Akama is concerned with *annotated logic programming*. Indeed, the starting point of annotated logics is paraconsistent logic programming, but the subject has been later expanded in various ways. Annotated logic programming can be considered as a tool for many applications. In this paper, they present several approaches to annotated logic programming.

Chapter 8 by Y. Kudo et al. reviews *rough set theory* in connection with modal logic. Rough set theory can serve as a basis for granularity computing and can be applied to deal with many problems in intelligent systems. It is well known that there are some connections between rough set theory and modal logic.

Chapter 9 by T. Murai et al. investigates paraconsistency and paracompleteness in Chellas's conditional logic using Scott–Montague semantics. It is possible to express inconsistency and incompleteness in conditional logic, and they provide several formal results.

Chapter 10 by F.A. Doria and C.A. Cosenza presents a logical approach to the so-called *efficient market* which means that stock prices fully reflect all available information in the market. They introduce the concept of almost efficient market and study its formal properties.

Chapter 11 by J.-M. Alliot et al. is about a logic called the *molecular interaction logic* to represent temporal reasoning in biological systems. The logic can semantically characterize *molecular interaction maps* (MIM) and formalize various reasoning on MIM.

Chapter 12 by S. Akama summarizes Abe's work on paraconsistent logics and their applications to engineering and surveys some of his projects shortly. The paper clarifies his ideas on paraconsistent engineering.

Most papers in this book are related to paraconsistent logics, and they tackle various problems by using paraconsistent logics. The book is dedicated to Jair Minoro Abe for his 60th birthday. I am grateful to contributors and referees.

Kawasaki, Japan   Seiki Akama
May 2016

Contents

Contributors

Jair Minoro Abe Research Group in Paraconsistent Logic Applications, UNISANTA, Santa Cecília University, Santos City, SP, Brazil; Graduate Program in Production Engineering, ICET, Paulista University, São Paulo, Brazil

Seiki Akama C-Republic, Asao-ku, Kawasaki, Japan

Jean-Marc Alliot INSERM/IRIT, University of Toulouse, Toulouse, France

Jean-Yves Beziau UFRJ—Federal University of Rio de Janeiro, Rio de Janeiro, Brazil; CNPq—Brazilian Research Council, Rio de Janeiro, Brazil

Carlos A. Cosenza Advanced Studies Research Group, HCTE, Fuzzy Sets Laboratory, Mathematical Economics Group, Production Engineering Program, COPPE, UFRJ, Rio Rj, Brazil

Newton C.A. da Costa Department of Philosophy, Federal University of Santa Catarina, Florianópolis, SC, Brazil

Clovis Misseno da Cruz Research Group in Paraconsistent Logic Applications, UNISANTA, Santa Cecília University, Santos City, SP, Brazil

João Inácio da Silva Filho Research Group in Paraconsistent Logic Applications, UNISANTA, Santa Cecília University, Santos City, SP, Brazil

Robert Demolombe INSERM/IRIT, University of Toulouse, Toulouse, France

Martín Diéguez INSERM/IRIT, University of Toulouse, Toulouse, France

Francisco Antonio Doria Advanced Studies Research Group, HCTE, Fuzzy Sets Laboratory, Mathematical Economics Group, Production Engineering Program, COPPE, UFRJ, Rio Rj, Brazil

Luis Fariñas del Cerro INSERM/IRIT, University of Toulouse, Toulouse, France

Gilles Favre INSERM/IRIT, University of Toulouse, Toulouse, France

Jean-Charles Faye INSERM/IRIT, University of Toulouse, Toulouse, France

Luís Fernando P. Ferrara Research Group in Paraconsistent Logic Applications, UNISANTA, Santa Cecília University, Santos City, SP, Brazil

Dorotéa Vilanova Garcia Research Group in Paraconsistent Logic Applications, UNISANTA, Santa Cecília University, Santos City, SP, Brazil

Décio Krause Department of Philosophy, Federal University of Santa Catarina, Florianópolis, SC, Brazil

Yasuo Kudo Muroran Institute of Technology, Muroran, Japan

Mauricio Conceição Mario Research Group in Paraconsistent Logic Applications, UNISANTA, Santa Cecília University, Santos City, SP, Brazil

Tetsuya Murai Chitose Institute of Science and Technology, Chitose, Japan

Kazumi Nakamatsu University of Hyogo, Himeji, Japan

Naji Obeid INSERM/IRIT, University of Toulouse, Toulouse, France

Alexandre Shozo Onuki Research Group in Paraconsistent Logic Applications, UNISANTA, Santa Cecília University, Santos City, SP, Brazil

Alexandre Rocco Research Group in Paraconsistent Logic Applications, UNISANTA, Santa Cecília University, Santos City, SP, Brazil

Olivier Sordet INSERM/IRIT, University of Toulouse, Toulouse, France

Chapter 1
Introduction

Seiki Akama

Dedicated to Jair Minoro Abe for his 60th birthday

Abstract Paraconsistent logic is a family of non-classical logics to tolerate inconsistency. Many systems of paraconsistent logics have been developed, and they are now applied to several areas including engineering. Jair Minoro Abe, who is an expert on annotated logics, is one of the important figures in paraconsistent logics. This book collects papers, addressing the importance of paraconsistent logics for several fields.

Keywords Paraconsistent logics · Non-classical logics · Annotated logics · J.M. Abe

1.1 Backgrounds

In the 1980s, I was working on logical foundations for intelligent systems. In particular, I was interested in automated reasoning and knowledge representation in Artificial Intelligence (AI). Unfortunately, some people in related areas believed that logic is of no use for the purpose. However, I believed that logic can serve as mathematical foundations for intelligent systems.

The main tool of logical approaches to AI was undoubtedly *classical logic*. Many researchers studied theorem-proving methods for classical logic, e.g. resolution, and tried to use it as a knowledge representation language. This means that proof theory

S. Akama (✉)
C-Republic, 1-20-1 Higashi-Yurigaoka, Asao-ku, Kawasaki 215-0012, Japan
e-mail: akama@jcom.home.ne.jp

© Springer International Publishing Switzerland 2016
S. Akama (ed.), *Towards Paraconsistent Engineering*, Intelligent Systems
Reference Library 110, DOI 10.1007/978-3-319-40418-9_1

1

can be applied to inference engine and model theory can be applied knowledge representation language. In the 1970s, logic programming languages like Prolog were developed.

The fact that classical logic was mainly considered in AI is not surprising since it is well-studied in the area of mathematical logic. But, classical logic has some limitations in the study of AI; see Minsky [6]. One of the serious difficulties is that it cannot deal with incomplete and inconsistent information. To overcome it, AI workers invented *non-monotonic logic* for common-sense reasoning.

Naturally, we may explore the use of *non-classical logic* in the study of AI. In the 1980s, this was not a defensible idea for AI. I studied several non-classical logics for AI. I started with modal logics and constructive logics, because these logics are suited to formalize incomplete information. But, I felt that the representation of inconsistent information is also important.

Based on the considerations I learned several paraconsistent logics. Unfortunately, I could not find intriguing applications to AI based on paraconsistent logics. I also found that there were some problems of the use of paraconsistent logics. One major problem is that to develop an automated reasoning method for paraconsistent logic is difficult. In addition, logical basis of paraconsistent logics is complicated. I will give a quick review of paraconsistent logics in Chap. 2.

In logic programming community, the representation of incomplete and inconsistent information in connection with common-sense reasoning is also regarded as an important problem. In 1987, Subramanian proposed an *annotated logic* for qualitative logic programming (cf. [7]) and *paraconsistent logic programming* (cf. [3]). I investigated these papers with great interest.

However, I was dissatisfied with the approach, since it is restrictive from a logical point of view. I believed that annotated logic can be formalized as a formal logical system and the work is intriguing. I tried to study the subject. In fact, it is important to explore foundations and applications of paraconsistent logics, although I was working on other research projects in the period.

In 1991, two important papers on annotated logic have been published; i.e., da Costa et al. [4, 5]. These papers in fact dealt with foundations for annotated logic. For me, the fact was shocking. In 1993, I invited Richard Sylvan in Japan, and he informed that Jair Minoro Abe wrote a dissertation on annotated logic; see Abe [1]. Unfortunately, Abe's dissertation was written in Portuguese. But I could suppose the results from the above two papers. Based on these papers, I stopped foundational work on annotated logic, but seeked a possibility of its applications to computer science.

In 1997, I attended the first World Congress on Paraconsistency held Ghent to present a paper on relevant counterfactuals. From the program, I knew that Abe presented several papers on annotated logic. I attended the session and questioned to him. After the session, I talked with him. It is not surprising that he could speak Japanese. I started the research collaboration with him. In Ghent, I also met Nakamatsu who studied annotated logic from the perspective on logic programming.

Since then, I worked with Abe and Nakamatsu on annotated logics and published many papers. Our goal was to established foundations and applications for annotated logic. We decided to write a monograph on annotated logic. In 2015, we published "Introduction to Annotated Logics" in 2015 by Springer; see Abe et al. [2].

To celebrate Abe's sixty birthday, I decided to edit a book for him. The project is important because he is one of the important figures in the area of paraconsistent logic in that he explored many applications of paraconsistent logics for engineering. I am happy to present this book to honor him and show progresses of paraconsistent logics.

Jair Minoro Abe was born in São Paulo, Brazil on October 6, 1955. He received bachelor and master degrees at Institute of Mathematics and Statics of University of São Paulo in 1978 and 1983, and doctor degree at Faculty of Philosophy, Letters and Human Sciences from University of São Paulo in 1992. He is now Full Professor of Paulista University. His research topics include non-classical logics and Artificial Intelligence. He is working on foundations and applications of paraconsistent logics, in particular, annotated logic. I will describe Abe's life and research in Chap. 12.

1.2 About This Book

The title of this book "Towards Paraconsistent Engineering" clearly describes Abe's research projects. Now, we summarize the contents of this book. Most papers are concerned with paraconsistent logics, and some papers addressed the usefulness of non-classical logics (Fig. 1.1).

Chapter 2 by S. Akama and N.C.A da Costa discusses the reason why paraconsistent logics are very useful to engineering. In fact, the use of paraconsistent logics

Fig. 1.1 Jair Minoro Abe

is a starting point of Abe's work. The ideas and history of paraconsistent logics are reviewed. The chapter is also useful for readers to read papers in this book.

Chapter 3 by N.C.A da Costa and D. Krause is concerned with an application of a paraconsistent logic to quantum physics. The paper reviews the authors' previous papers on the concept of complementarity introduced by Bohr. Logical foundations for quantum mechanics have been worked out so far. The paper reveals that the authors' paraconsistent logic can serve as the basis for the important problem in quantum mechanics.

Chapter 4 by J.-Y. Beziau proposes two three-valued paraconsistent logics, which are 'genuine' in the sense that they obey neither $p, \neg p \vdash q$ nor $\vdash \neg(p \wedge \neg p)$. Beziau investigates their properties and relations to other paraconsistent logics. His work is seen as a new approach to three-valued paraconsistent logics.

Chapter 5 by S. Akama surveys *annotated logics* which have been developed as paraconsistent and paracomplete logics by Abe and others. The paper presents the formal and practical aspects of annotated logics and suggests their further applications for paraconsistent engineering.

Chapter 6 by J.I. da Silva Filho et al. discusses an application of the annotated logic called *PAL2v* based on two truth-values for *Paraconsistent Artificial Neural Network* (PANet), showing an algorithmic structure for handling actual problems. The paper is one of the interesting engineering applications of paraconsistent logic.

Chapter 7 by K. Nakamatsu and S. Akama is concerned with *annotated logic programming*. Indeed the starting point of annotated logics is paraconsistent logic programming, but the subject has been later expanded in various ways. Annotated logic programming can be considered as a tool for many applications. In this paper, they present several approaches to annotated logic programming.

Chapter 8 by Y. Kudo et al. reviews *rough set theory* in connection with modal logic. Rough set theory can serve as a basis for granularity computing and can be applied to deal with many problems in intelligent systems. It is well known that there are some connections between rough set theory and modal logic.

Chapter 9 by T. Murai et al. investigates paraconsistency and paracompleteness in Chellas's conditional logic using Scott-Montague semantics. It is possible to express inconsistency and incompleteness in conditional logic, and they provide several formal results.

Chapter 10 by F.A. Doria and C.A. Cosenza presents a logical approach to the so-called *efficient market* which means that stock prices fully reflect all available information in the market. They introduce the concept of almost efficient market and study its formal properties.

Chapter 11 by J.-M. Alliot et al. is about a logic called the *Molecular Interaction Logic* to represent temporal reasoning in biological systems. The logic can semantically characterize *Molecular Interaction Maps* (MIM) and formalize various reasoning on MIM.

Chapter 12 by S. Akama summarizes Abe's work on paraconsistent logics and their applications to engineering, and surveys some of his projects shortly. The paper clarifies his ideas on paraconsistent engineering.

Acknowledgments I am grateful to Prof. Abe for his comments.

References

1. Abe, J.M.: On the Foundations of Annotated Logics (in Portuguese), Ph.D. Thesis, University of São Paulo, Brazil (1992)
2. Abe, J.M., Akama, S., Nakamatsu, K.: Introduction to Annotated Logics. Springer, Heidelberg (2016)
3. Blair, H.A., Subrahmanian, V.S.: Paraconsistent logic programming. Theor. Comput. Sci. **68**, 135–154 (1989)
4. da Costa, N.C.A., Abe, J.M., Subrahmanian, V.S.: Remarks on annotated logic. Zeitschrift für mathematische Logik und Grundlagen der Mathematik **37**, 561–570 (1991)
5. da Costa, N.C.A., Subrahmanian, V.S., Vago, C.: The paraconsistent logic PT. Zeitschrift für mathematische Logik und Grundlagen der Mathematik **37**, 139–148 (1991)
6. Minsky, M.: A framework for representing knowledge. In: Haugeland, J. (ed.) Mind-Design, pp. 95–128. MIT Press, Cambridge (1975)
7. Subrahmanian, V.: On the semantics of quantitative logic programs. In: Proceedings of the 4th IEEE Symposium on Logic Programming, pp. 173–182 (1987)

Chapter 2
Why Paraconsistent Logics?

Seiki Akama and Newton C.A. da Costa

Dedicated to Jair Minoro Abe for his 60th birthday.

Abstract In this chapter, we briefly review paraconsistent logics which are closely related to the topics in this book. We give an exposition of their history and formal aspects. We also address the importance of applications of paraconsistent logics to engineering.

Keywords Paraconsistent logics · Contradiction inconsistency · Paraconsistency

2.1 Introduction

Paraconsistent logic is a logical system for inconsistent but non-trivial formal theories. It is classified as *non-classical logic* in the sense that it can be employed as a rival to classical logic. Paraconsistent logic has many applications and it can serve as a foundation for engineering because some engineering problems must solve inconsistent information. However standard classical logic cannot tolerate it. In this regard, paraconsistent is promising.

Here, we give a quick review of paraconsistent logic that is helpful to the reader. Let T be a theory whose underlying logic is L. T is called *inconsistent* when it contains theorems of the form A and $\neg A$ (the negation of A), i.e.,

S. Akama (✉)
C-Republic, 1-20-1 Higashi-Yurigaoka, Asao-ku, Kawasaki 215-0012, Japan
e-mail: akama@jcom.home.ne.jp

N.C.A. da Costa
Department of Philosophy, Federal University of Santa Catarina,
Florianópolis, SC, Brazil
e-mail: ncacosta@terra.com.br

© Springer International Publishing Switzerland 2016
S. Akama (ed.), *Towards Paraconsistent Engineering*, Intelligent Systems
Reference Library 110, DOI 10.1007/978-3-319-40418-9_2

$$T \vdash_L A \text{ and } T \vdash_L \neg A$$

where $\vdash_L$ denotes the provability relation in L. If T is not inconsistent, it is called *consistent*.

T is said to be *trivial*, if all formulas of the language are also theorems of T. Otherwise, T is called *non-trivial*. Then, for trivial theory T, $T \vdash_L B$ for any formula B. Note that trivial theory is not interesting since every formula is provable.

If L is classical logic (or one of several others, such as intuitionistic logic), the notions of inconsistency and triviality agree in the sense that T is inconsistent iff T is trivial. So, in trivial theories the extensions of the concepts of formula and theorem coincide.

A *paraconsistent logic* is a logic that can be used as the basis for inconsistent but non-trivial theories. In this regard, sentences of paraconsistent theories do not satisfy, in general, the *principle of non-contradiction*, i.e., $\neg(A \wedge \neg A)$.[1]

Similarly, we can define the notions of paracomplete logic and theory. A *paracomplete logic* is a logic, in which the *principle of excluded middle*, i.e., $A \vee \neg A$ is not a theorem of that logic. In this sense, intuitionistic logic is one of the paracomplete logics. A *paracomplete theory* is a theory based on paracomplete logic.

Finally, a logic which is simultaneously paraconsistent and paracomplete is called *non-alethic logic*.

The structure of this paper is as follows. In Sect. 2.2, we describe the history of paraconsistent logic. In Sect. 2.3, major approaches to paraconsistent logic are given with formal descriptions. In Sect. 2.4, other paraconsistent logics are briefly reviewed.

2.2 History

This section surveys the history of paraconsistent logic. Paraconsistent logics have recently proved attracted to many people, but they have a longer history than classical logic. For example, Aristotle developed a logical theory that can be interpreted to be paraconsistent. But, paraconsistent logics in the modern sense were formally devised in the 1950s.

In 1910, the Russian logician Nikolaj A. Vasil'ev (1880–1940) and the Polish logician Jan Łukasiewicz (1878–1956) independently glimpsed the possibility of developing paraconsistent logics. Vasil'ev's *imaginary logic* can be seen as a paraconsistent reformulation of Aristotle's *syllogistic*; see Vasil'ev [54].

It was here pointed out that Łukasiewicz's *three-valued logic* is a forerunner of the many-valued approach to paraconsistency, although he did not explicitly discuss paraconsistency; see Łukasiewicz [43].

---

[1] In fact, in some systems of paraconsistent logic, like da Costa's systems C_n, the "good" propositions do satisfy this principle.

However, we believe that the history of paraconsistent logic started in 1948. Stanislaw Jaśkowski (1896–1965) proposed a paraconsistent propositional logic, now called *discursive logic* (or discussive logic) in 1948; see Jaśkowski [37, 38]. Discursive logic is based on modal logic, and it is classified as the modal approach to paraconsistency.

Independently, some years later, the Brazilian logician Newton C.A. da Costa (1929-) constructed for the first time hierarchies of paraconsistent propositional calculi $C_i(1 \leq i \leq \omega)$ and its first-order and higher-order extensions; see da Costa [28]. da Costa's logics are called the *C-system*, which is based on the non-standard interpretation of negation which is dual to intuitionistic negation.

A different route to paraconsistent logic may be found in the so-called *relevance logic* (or relevant logic), which was originally developed by Anderson and Belnap in the 1960s; see Anderson and Belnap [11] and Anderson, Belnap and Dunn [12]. Anderson and Belnap's approach addresses a correct interpretation of implication $A \rightarrow B$, in which A and B should have some connection. Its semantic interpretation raises the issues of paraconsistency, and some (not all) relevance logics are in fact paraconsistent.

The above three approaches are considered the major approaches to paraconsistent logics, many paraconsistent logics have been proposed in the literature. They have been developed from some motivation.

2.3 Approaches to Paraconsistent Logic

The section formally reviews several paraconsistent logics, restricting to the principal paraconsistent logics. But, it is far from complete, and the reader should consult in-depth exposition in the relevant reference.

We can list the three logics as the major approaches:

- Discursive logic
- C-systems
- Relevant (relevance) logic

Discursive logic, also known as discussive logic, was proposed by Jaśkowski [37, 38], which is regarded as a non-adjunctive approach. *Adjunction* is a rule of inference of the form: from $\vdash A$ and $\vdash B$ to $\vdash A \wedge B$. Discursive logic can avoid explosion by prohibiting adjunction.

It was a formal system J satisfying the conditions: (a) from two contradictory propositions, it should not be possible to deduce any proposition; (b) most of the classical theses compatible with (a) should be valid; (c) J should have an intuitive interpretation.

Such a calculus has, among others, the following intuitive properties remarked by Jaśkowski himself: suppose that one desires to systematize in only one deductive system all theses defended in a discussion. In general, the participants do not confer the same meaning to some of the symbols.

One would have then as theses of a deductive system that formalize such a discussion, an assertion and its negation, so both are "true" since it has a variation in the sense given to the symbols. It is thus possible to regard discursive logic as one of the so-called *paraconsistent logics*.

Jaśkowski's D_2 contains propositional formulas built from logical symbols of classical logic. In addition, the possibility operator $\diamond$ in S5 is added. Based on the possibility operator, three discursive logical symbols can be defined as follows:

discursive implication: $A \rightarrow_d B =_{def} \diamond A \rightarrow B$
discursive conjunction: $A \wedge_d B =_{def} \diamond A \wedge B$
discursive equivalence: $A \leftrightarrow_d B =_{def} (A \rightarrow_d B) \wedge_d (B \rightarrow_d A)$

Additionally, we can define discursive negation $\neg_d A$ as $A \rightarrow_d false$. Jaśkowski's original formulation of D_2 in [38] used the logical symbols: $\rightarrow_d$, $\leftrightarrow_d$, $\vee$, $\wedge$, $\neg$, and he later defined $\wedge_d$ in [38].

The following axiomatization due to Kotas [42] has the following axioms and the rules of inference.

Axioms
(A1) $\square(A \rightarrow (\neg A \rightarrow B))$
(A2) $\square((A \rightarrow B) \rightarrow ((B \rightarrow C) \rightarrow (A \rightarrow C)))$
(A3) $\square((\neg A \rightarrow A) \rightarrow A)$
(A4) $\square(\square A \rightarrow A)$
(A5) $\square(\square(A \rightarrow B) \rightarrow (\square A \rightarrow \square B))$
(A6) $\square(\neg \square A \rightarrow \square \neg \square A)$

Rules of Inference
(R1) substitution rule
(R2) $\square A, \square(A \rightarrow B)/\square B$
(R3) $\square A/\square\square A$
(R4) $\square A/A$
(R5) $\neg\square\neg\square A/A$

There are other axiomatizations of D_2, but we omit the details here. Discursive logics are considered weak as a paraconsistent logic, but they have some applications, e.g. logics for vagueness.

C-systems are paraconsistent logics due to da Costa which can be a basis for inconsistent but non-trivial theories; see da Costa [28]. The important feature of da Costa systems is to use novel interpretation, which is non-truth-functional, of negation avoiding triviality.

Here, we review C-system C_1 due to da Costa [28]. The language of C_1 is based on the logical symbols: $\wedge$, $\vee$, $\rightarrow$, and $\neg$. $\leftrightarrow$ is defined as usual. In addition, a formula A°, which is read "A is well-behaved", is shorthand for $\neg(A \wedge \neg A)$. The basic ideas of C_1 contain the following: (1) most valid formulas in the classical logic hold, (2) the law of non-contradiction $\neg(A \wedge \neg A)$ should not be valid, (3) from two contradictory formulas it should not be possible to deduce any formula.

The Hilbert system of C_1 extends the positive intuitionistic logic with the axioms for negation.

da Costa's C_1

Axioms

(DC1) $A \rightarrow (B \rightarrow A)$

(DC2) $(A \rightarrow B) \rightarrow ((A \rightarrow (B \rightarrow C)) \rightarrow (A \rightarrow C))$

(DC3) $(A \wedge B) \rightarrow A$

(DC4) $(A \wedge B) \rightarrow B$

(DC5) $A \rightarrow (B \rightarrow (A \wedge B))$

(DC6) $A \rightarrow (A \vee B)$

(DC7) $B \rightarrow (A \vee B)$

(DC8) $(A \rightarrow C) \rightarrow ((B \rightarrow C) \rightarrow ((A \vee B) \rightarrow C))$

(DC9) $B° \rightarrow ((A \rightarrow B) \rightarrow ((A \rightarrow \neg B) \rightarrow \neg A))$

(DC10) $(A° \wedge B°) \rightarrow (A \wedge B)° \wedge (A \vee B)° \wedge (A \rightarrow B)°$

(DC11) $A \vee \neg A$

(DC12) $\neg\neg A \rightarrow A$

Rules of Inference

(MP) $\vdash A,\ \vdash A \rightarrow B \Rightarrow \vdash B$

Here, (DC1)–(DC8) are axioms of the positive intuitionistic logic. (DC9) and (DC10) play a role for the formalization of paraconsistency.

A semantics for C_1 can be given by a two-valued valuation; see da Costa and Alves [29]. We denote by $\mathcal{F}$ the set of formulas of C_1. A valuation is a mapping v from $\mathcal{F}$ to $\{0, 1\}$ satisfying the following:

$v(A) = 0 \Rightarrow v(\neg A) = 1$

$v(\neg\neg A) = 1 \Rightarrow v(A) = 1$

$v(B°) = v(A \rightarrow B) = v(A \rightarrow \neg B) = 1 \Rightarrow v(A) = 0$

$v(A \rightarrow B) = 1 \Leftrightarrow v(A) = 0$ or $v(B) = 1$

$v(A \wedge B) = 1 \Leftrightarrow v(A) = v(B) = 1$

$v(A \vee B) = 1 \Leftrightarrow v(A) = 1$ or $v(B) = 1$

$v(A°) = v(B°) = 1 \Rightarrow v((A \wedge B)°) = v((A \vee B)°) = v((A \rightarrow B)°) = 1$

Note here that the interpretations of negation and double negation are not given by biconditional. A formula A is *valid*, written $\models A$, if $v(A) = 1$ for every valuation v. Completeness holds for C_1. It can be shown that C_1 is complete for the above semantics.

Da Costa system C_1 can be extended to C_n ($1 \le n \le \omega$). Now, A^1 stands for $A°$ and A^n stands for $A^{n-1} \wedge (A^{(n-1)})°$, $1 \le n \le \omega$.

Then, da Costa system C_n ($1 \le n \le \omega$) can be obtained by (DC1)–(DC8), (DC12), (DC13) and the following:

(DC9n) $B^{(n)} \rightarrow ((A \rightarrow B) \rightarrow ((A \rightarrow \neg B) \rightarrow \neg A))$

(DC10n) $(A^{(n)} \wedge B^{(n)}) \rightarrow (A \wedge B)^{(n)} \wedge (A \vee B)^{(n)} \wedge (A \rightarrow B)^{(n)}$

Note that the da Costa system C_ω has the axioms (DC1)–(DC8), (DC12) and (DC13). Later, da Costa investigated first-order and higher-order extensions of C-systems.

Relevance logic, also called *relevant logic* is a family of logics based on the notion of relevance in conditionals. Historically, relevance logic was developed to avoid the *paradox of implications*; see Anderson and Belnap [11, 12].

Anderson and Belnap formalized a relevant logic R to realize a major motivation, in which they do not admit $A \to (B \to A)$. Later, various relevance logics have been proposed. Note that not all relevance logics are paraconsistent but some are considered important as paraconsistent logics.

Routley and Meyer proposed a basic relevant logic B, which is a minimal system having the so-called *Routley-Meyer semantics*. Thus, B is an important system and we review it below; see Routley et al. [51].

The language of B contains logical symbols: $\sim$, &, $\vee$ and $\to$ (relevant implication). A Hilbert system for B is as follows:

Relevant Logic B
Axioms
(BA1) $A \to A$
(BA2) $(A\&B) \to A$
(BA3) $(A\&B) \to B$
(BA4) $((A \to B)\&(A \to C)) \to (A \to (B\&C))$
(BA5) $A \to (A \vee B)$
(BA6) $B \to (A \vee B)$
(BA7) $(A \to C)\&(B \to C)) \to ((A \vee B) \to C)$
(BA8) $(A\&(B \vee C)) \to (A\&B) \vee C)$
(BA9) $\sim\sim A \to A$
Rules of Inference
(BR1) $\vdash A, \vdash A \to B \Rightarrow \vdash B$
(BR2) $\vdash A, \vdash B \Rightarrow \vdash A\&B$
(BR3) $\vdash A \to B, \vdash C \to D \Rightarrow \vdash (B \to C) \to (A \to D)$
(BR4) $\vdash A \to \sim B \Rightarrow \vdash B \to \sim A$

A Hilbert system for Anderson and Belnap's R is as follows:

Relevance Logic R
Axioms
(RA1) $A \to A$
(RA2) $(A \to B) \to ((C \to A) \to C \to B))$
(RA3) $(A \to (A \to B)) \to (A \to B)$
(RA4) $(A \to (B \to C)) \to (B \to (A \to C))$
(RA5) $(A\&B) \to A$
(RA6) $(A\&B) \to B$
(RA7) $((A \to B)\&(A \to C)) \to (A \to (B\&C))$
(RA8) $A \to (A \vee B)$
(RA9) $B \to (A \vee B)$
(RA10) $((A \to C)\&(B \vee C)) \to ((A \vee B) \to C))$
(RA11) $(A\&(B \vee C)) \to ((A\&B) \vee C)$
(RA12) $(A \to \sim A) \to \sim A$

(RA13) $(A \rightarrow \sim B)) \rightarrow (B \rightarrow \sim A)$
(RA14) $\sim\sim A \rightarrow A$

Rules of Inference

(RR1) $\vdash A, \vdash A \rightarrow B \Rightarrow \vdash B$
(RR2) $\vdash A, \vdash B \Rightarrow \vdash A\&B$

Routley et al. considered some axioms of R are too strong and formalized rules instead of axioms. Notice that B is a paraconsistent but R is not.

Next, we give a Routley-Meyer semantics for B. A model structure is a tuple $\mathcal{M} = \langle K, N, R, *, v \rangle$, where K is a non-empty set of worlds, $N \subseteq K$, $R \subseteq K^3$ is a ternary relation on K, $*$ is a unary operation on K, and v is a valuation function from a set of worlds and a set of propositional variables $\mathcal{P}$ to $\{0, 1\}$.

There are some restrictions on. v satisfies the condition that $a \leq b$ and $v(a, p)$ imply $v(b, p) = 1$ for any $a, b \in K$ and any $p \in \mathcal{P}$. $a \leq b$ is a pre-order relation defined by $\exists x(x \in N$ and $Rxab)$. The operation $*$ satisfies the condition $a^{**} = a$.

For any propositional variable p, the truth condition $\models$ is defined: $a \models p$ iff $v(a, p) = 1$. Here, $a \models p$ reads "p is true at a". $\models$ can be extended for any formulas in the following way:

$a \models \sim A \Leftrightarrow a^* \not\models A$
$a \models A\&B \Leftrightarrow a \models A$ and $a \models B$
$a \models A \vee B \Leftrightarrow a \models A$ or $a \models B$
$a \models A \rightarrow B \Leftrightarrow \forall bc \in K(Rabc$ and $b \models A \Rightarrow c \models B)$

A formula A is *true* at a in $\mathcal{M}$ iff $a \models A$. A is *valid*, written $\models A$, iff A is true on all members of N in all model structures.

Routley et al. provides the completeness theorem for B with respect to the above semantics using canonical models; see [51].

A model structure for R needs the following conditions.

$R0aa$
$Rabc \Rightarrow Rbac$
$R^2(ab)cd \Rightarrow R^2a(bc)d$ $Raaa$
$a^{**} = a$
$Rabc \Rightarrow Rac^*b^*$
$Rabc \Rightarrow a' \leq a \Rightarrow Ra'bc$

where R^2abcd is shorthand for $\exists x(Raxd$ and $Rxcd)$. The completeness theorem for the Routley-Meyer semantics can be proved for R; see [11, 12].

The reader is advised to consult Anderson and Belnap [11], Anderson et al. [12], and Routley et al. [51] for details. A more concise survey on the subject may be found in Dunn [32].

Belnap proposed a famous *four-valued logic* in Belnap [21, 22], which is closely related to relevant logic and paraconsistent logic. Belnap's four-valued logic aims to formalize the internal states of a computer.

There are four states, i.e. (*T*), (*F*), (*None*) and (*Both*), to recognize an input in a computer. Based on these states, a computer can compute the following suitable outputs.

(*T*)   a proposition is true.
(*F*)   a proposition is false.
(*N*)   a proposition is neither true nor false.
(*B*)   a proposition is both true and false.

Here, (*N*) and (*B*) abbreviate (*None*) and (*Both*), respectively. From the above,
(*N*) corresponds to incompleteness and (*B*) inconsistency. Four-valued logic can be
thus seen as a natural extension of three-valued logic. In fact, Belnap's four-valued
logic can model both incomplete information (*N*) and inconsistent information (*B*).

Belnap proposed two four-valued logics **A4** and **L4**. The former can cope only
with atomic formulas, whereas the latter can handle compound formulas. **A4** is based
on the *approximation lattice*, which is shown in Fig. 2.1.

Here, *B* is the least upper bound and *N* is the greatest lower bound with respect
to the ordering $\leq$. Observe that in the lattice *FOUR* in Fig. 2.1, we used $t, f, \bot, \top$
instead of T, F, N, B, respectively.

The logic **L4** has logical symbols; $\sim, \wedge, \vee$. Its truth-values is $4 = \{T, F, N, B\}$
with a different ordering. The lattice **L4** is shown in Fig. 2.2.

One of the features of **L4** is the monotonicity of logical symbols. Let f be a logical
operation. It is said that f is monotonic iff $a \subseteq b \Rightarrow f(a) \subseteq f(b)$. To guarantee the
monotonicity of conjunction and disjunction, they must satisfy the following:

$$a \wedge b = a \Leftrightarrow a \vee b = b$$
$$a \wedge b = b \Leftrightarrow a \vee b = a$$

Logical symbols in **L4** obey th truth-value tables in Table 2.1.

Fig. 2.1 Approximation
lattice **L4**

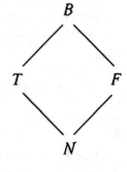

Fig. 2.2 Logical lattice **L4**

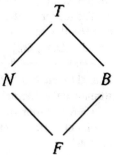

Table 2.1 Truth-value tables of **L4**

	N	F	T	B
~	B	T	F	N

∧	N	F	T	B
N	N	F	N	F
F	F	F	F	F
T	N	F	T	B
B	F	F	B	B

∨	N	F	T	B
N	N	N	T	T
F	N	F	T	B
T	T	T	T	T
B	T	B	T	B

Belnap gave a semantics for the language with the above logical symbols. A *setup* is a mapping a set of atomic formulas *Atom* to the set **4**. Then, formulas of **L4** are defined as follows:

$s(A \wedge B) = s(A) \wedge s(B)$

$s(A \vee B) = s(A) \vee s(B)$

$s(\sim A) = \sim s(A)$

Further, Belnap defined an entailment relation → as follows:

$A \rightarrow B \Leftrightarrow s(A) \leq s(B)$

for all setups s. Note that → is not a logical connective for implication but an entailment relation. The entailment relation → can be axiomatized as follows:

$(A_1 \wedge ... \wedge A_m) \rightarrow (B_1 \vee ... \vee B_n)$ (A_i shares some B_j)

$(A \vee B) \rightarrow C \Leftrightarrow (A \rightarrow C)$ and $(B \rightarrow C)$

$A \rightarrow B \ \Leftrightarrow \sim B \rightarrow \sim A$

$A \vee B \leftrightarrow B \vee A, \ A \wedge B \leftrightarrow B \wedge A$

$A \vee (B \vee C) \leftrightarrow (A \vee B) \vee C$

$A \wedge (B \wedge C) \leftrightarrow (A \wedge B) \wedge C$

$A \wedge (B \vee C) \leftrightarrow (A \wedge B) \vee (A \wedge C)$

$A \vee (B \wedge C) \leftrightarrow (A \vee B) \wedge (A \vee C)$

$(B \vee C) \wedge A \leftrightarrow (B \wedge A) \vee (C \wedge A)$

$(B \wedge C) \vee A \leftrightarrow (B \vee A) \wedge (C \vee A)$

$\sim\sim A \leftrightarrow A$

$\sim (A \wedge B) \leftrightarrow \sim A \vee \sim B, \ \sim (A \vee B) \leftrightarrow \sim A \wedge \sim B$

$A \rightarrow B, B \rightarrow C \ \Leftrightarrow A \rightarrow C$

$A \leftrightarrow B, B \leftrightarrow C \ \Leftrightarrow A \leftrightarrow C$

$A \rightarrow B \ \Leftrightarrow A \leftrightarrow (A \wedge B) \ \Leftrightarrow (A \vee B) \leftrightarrow B$

Note here that $(A \wedge \sim A) \rightarrow B$ and $A \rightarrow (B \vee \sim B)$ cannot be derived in this axiomatization. It can be shown that the logic given above is shown to be equivalent to the system of *tautological entailment*; see [11, 12].

An alternative semantics for tautological entailment based on the notion of fact was worked out by van Fraassen [53]. Belnap's **A4** is used as one of the lattice of truth-values as *FOUR*. In this regard, Belnap's four-valued logic is considered as the important background on annotated logics.

2.4 Other Paraconsistent Logics

Although the above three logics are famous approaches to paraconsistent logics, there is a rich literature on paraconsistent logics. Arruda [15] reviewed a survey on paraconsistent logics, and Priest et al. [49] contains interesting papers on paraconsistent logics in the 1980s. For a recent survey, we refer Priest [47]. We can also find a Handbook surveying various subjects related to paraconsistency by Beziau et al. [23].

In 1997, The First World Congress on Paraconsistency (WCP'1997) was held at the University of Ghent, Belgium; see Batens et al. [20]. The Second World Congress on Paraconsistency (WCP'200) was held at Juquehy-Sao Paulo, Brazil; see Carnielli et al. [26].

In the 1990s paraconsistent logics became one of the major topics in logic in connection with other areas, in particular, computer science. Below we review some of those systems of paraconsistent logics.

The modern history of paraconsistent logic started with Vasil'ev's *imaginary logic*. In 1910, Vasil'ev proposed an extension of Aristotle's syllogistic allowing the statement of the form S is both P and not-P; see Vasil'ev [54].

Thus, imaginary logic can be viewed as a paraconsistent logic. Unfortunately, little work has been done on focusing on its formalization from the viewpoint of modern logic. A survey of imaginary logic can be found in Arruda [15].

In 1954, Asenjo developed a calculus of antinomies in his dissertation; see Asenjo [16]. Asenjo's work was published before da Costa's work, but it seems that Asenjo's approach has been neglected. Asenjo's idea is to interpret the truth-value of *antinomy* as both true and false using Kleene's strong three-valued logic.

His proposed calculus is non-trivially inconsistent propositional logic, whose axiomatization can be obtained from Kleene's [39] axiomatization of classical propositional logic by deleting the axiom $(A \to B) \to ((A \to \neg B) \to \neg A)$.

In constructivism, an idea of constructing paraconsistent logics may be found. In 1949, Nelson [44] proposed a *constructive logic with strong negation* as an alternative to intuitionistic logic, in which *strong negation* (or constructible negation) is introduced to improve some weaknesses of intuitionistic negation.

Constructive logic N extends positive intuitionistic logic Int^+ with the following axioms for *strong negation* $\sim$:

(N1) $(A \wedge \sim A) \to B$
(N2) $\sim\sim A \leftrightarrow A$
(N3) $\sim (A \to B) \leftrightarrow (A \wedge \sim B)$
(N4) $\sim (A \wedge B) \leftrightarrow (\sim A \vee \sim B)$
(N5) $\sim (A \vee B) \leftrightarrow (\sim A \wedge \sim B)$

In N, intuitionistic negation $\neg$ can be defined as $\neg A \leftrightarrow A \to (B \wedge \sim B)$. If we delete (N1) from N, we can obtain a paraconsistent constructive logic N^- of Almukdad and Nelson [10]. Akama [3–8] extensively studied Nelson's constructive logics with strong negation; also see Wansing [55].

Table 2.2 Truth-value tables of Kleene's strong three-valued logic

A	$\neg A$
T	F
I	I
F	T

A	B	$A \wedge B$	$A \vee B$	$A \rightarrow_K B$
T	T	T	T	T
T	F	F	T	F
T	I	I	T	I
F	T	F	F	T
F	F	F	F	T
F	I	F	I	T
I	T	I	T	T
I	F	F	I	I
I	I	I	I	I

In 1959, Nelson [45] developed a constructive logic S which lacks contraction $(A \rightarrow (A \rightarrow B)) \rightarrow (A \rightarrow B)$ and discussed its aspects as a paraconsistent logic. Akama [7] gave a detailed presentation of Nelson's paraconsistent constructive logics. Akama et al. [9] proposed a constructive discursive logic based on Nelson's constructive logic.

In 1979, Priest [46] proposed a *logic of paradox*, denoted *LP*, to deal with the semantic paradox. The logic is of special importance to the area of paraconsistent logics. *LP* can be semantically defined by Kleene's strong three-valued logic whose truth-value tables are as Table 2.2.

Here, T and F denote truth and falsity, and the third truth-value I reads "undefined"; see Kleene [39].

Łukasiewicz's three-valued logic is interpreted by the above truth-value tables of Kleene's three-valued logic except for implication. Let $\rightarrow_L$ be the implication in Łukasiewicz's three-valued logic. Then, the truth-value tables are described as Table 2.3.

Here, the third truth-value reads "possible"; see Łukasiewicz [43]. Kleene's three-valued logic was used as a basis for reasoning about incomplete information in computer science.

Priest re-interpreted the truth-value tables of Kleene's strong three-valued logic, namely read the third-truth value as both true and false (B) rather than neither true nor false (I), and assumed that (T) and (B) are designated values. The idea has already been considered in Asenjo [16] and Belnap [21, 22].

Consequently, ECQ: $A, \sim A \models B$ is invalid. Thus, *LP* can be seen as a paraconsistent logic. Unfortunately, (material) implication in *LP* does not satisfy *modus ponens*. It is, however, possible to introduce relevant implications as real implication into *LP*.

Table 2.3 Truth-value tables of Łukasiewicz's three-valued logic

A	$\sim A$
T	F
I	I
F	T

A	B	$A \wedge B$	$A \vee B$	$A \to_L B$
T	T	T	T	T
T	F	F	T	F
T	I	I	T	I
F	T	F	F	T
F	F	F	F	T
F	I	F	I	T
I	T	I	T	T
I	F	F	I	I
I	I	I	I	T

Priest developed a semantics for *LP* by means of a truth-value assignment relation rather than a truth-value assignment function. Let $\mathcal{P}$ be the set of propositional variables. Then, an evaluation η is a subset of $\mathcal{P} \times \{0, 1\}$. A proposition may only relate to 1 (true), it may only relate to 0 (false), it may relate to both 1 and 0 or it may relate to neither 1 nor 0. The evaluation is extended to a relation for all formulas as follows:

$\neg A\eta 1$ iff $A\eta 0$
$\neg A\eta 0$ iff $A\eta 1$
$A \wedge B\eta 1$ iff $A\eta 1$ and $B\eta 1$
$A \wedge B\eta 0$ iff $A\eta 0$ or $B\eta 0$
$A \vee B\eta 1$ iff $A\eta 1$ or $B\eta 1$
$A \vee B\eta 0$ iff $A\eta 0$ and $B\eta 0$

If we define validity in terms of truth preservation under all relational evaluations, then we obtain *first-degree entailment* which is a fragment of relevance logics.

Using *LP*, Priest advanced his research program to tackle various philosophical and logical issues; see Priest [47, 48] for details. For instance, in *LP*, the liar sentence can be interpreted as both true and false. It is also observed that Priest promoted the philosophical view called *dialetheism* which claims that there are true contradictions. In fact, dialetheism has been extensively discussed by many people.

Since the beginning of the 1990s, Batens developed the so-called *adaptative logics* in Batens [18, 19]. These logics are considered as improvements of *dynamic dialectical logics* investigated in Batens [17]. *Inconsistency-adaptive logics* as developed by Batens [18] can serve as foundations for paraconsistent and non-monotonic logics.

Adaptive logics formalized classical logic as "dynamic logic". Here, "dynamic logic" is not the family of logics with the same name studied in computer science. A logic is *adaptive* iff it adapts itself to the specific premises to which it is applied. In this sense, adaptive logics can model the dynamics of human reasoning. There are two sorts of dynamics, i.e., *external dynamics* and *internal dynamics*.

The external dynamics is stated as follows. If new premises become available, then consequences derived from the earlier premise set may be withdrawn. In other words, the external dynamics results from the *non-monotonic* character of the consequence relations.

Let $\vdash$ be a consequence relation, Γ, Δ be sets of formulas, and A be a formula. Then, the external dynamics is formally presented as: $\Gamma \vdash A$ but $\Gamma \cup \Delta \nvdash A$ for some Γ, Δ and A. In fact, the external dynamics is closely related to the notion of *non-monotonic reasoning* in AI.

The internal dynamics is very different from the external one. Even if the premise set is constant, certain formulas are considered as derived at some stage of the reasoning process, but are considered as not derived at a later stage. For any consequence relation, insight in the premises is gained by deriving consequences from them.

In the absence of a positive test, this results in the internal dynamics. Namely, in the internal dynamics, reasoning has to adapt itself by withdrawing an application of the previously used inference rule, if we infer a contradiction at a later stage. Adaptive logics are logics based on the internal dynamics.

An Adaptive Logic *AL* can be characterized as a triple:

(i) A *lower limit logic* (LLL)
(ii) A set of *abnormalities*
(iii) An *adaptive strategy*

The lower limit logic *LLL* is any monotonic logic, e.g., classical logic, which is the stable part of the adaptive logic. Thus, *LLL* is not subject to adaptation. The set of abnormalities Ω comprises the formulas that are presupposed to be false, unless and until proven otherwise.

In many adaptive logics, Ω is the set of formulas of the form $A \wedge \sim A$. An adaptive strategy specifies a strategy of the applications of inference rules based on the set of abnormalities.

If the lower limit logic *LLL* is extended with the requirement that no abnormality is logically possible, one obtains a monotonic logic, which is called the *upper limit logic ULL*. Semantically, an adequate semantics for the upper limit logic can be obtained by selecting that lower limit logic models that verify no abnormality.

The name "abnormality" refers to the upper limit logic. *ULL* requires premise sets to be normal, and 'explodes' abnormal premise sets (assigns them the trivial consequence set).

If the lower limit logic is classical logic *CL* and the set of abnormalities comprises formulas of the form $\exists A \wedge \exists \sim A$, then the upper limit logic obtained by adding to *CL* the axioms $\exists A \rightarrow \forall A$. If, as is the case for many inconsistency-adaptive logics, the lower limit logic is a paraconsistent logic *PL* which contains *CL*, and the set of abnormalities comprises the formulas of the form $\exists(A \wedge \sim A)$, then the upper limit logic is *CL*. The adaptive logics interpret the set of premises 'as much as possible' in agreement with the upper limit logic; it avoids abnormalities 'in as far as the premises permit'.

Adaptive logics provide a new way of thinking of the formalization of paraconsistent logics in view of the dynamics of reasoning. Although inconsistency-adaptive logic is paraconsistent logic, applications of adaptive logics are not limited to paraconsistency. From a formal point of view, we can count adaptive logics as promising paraconsistent logics.

However, for applications, we may face several obstacles in automating reasoning in adaptive logics in that proofs in adaptive logics are dynamic with a certain adaptive strategy. Thus, the implementation is not easy, and we have to choose an appropriate adaptive strategy depending on applications.

Carnelli proposed the *Logics of Formal Inconsistency* (LFI), which are logical systems that treat consistency and inconsistency as mathematical objects; see Carnelli et al. [27]. One of the distinguishing features of these logics is that they can internalize the notions of consistency and inconsistency at the object-level.

And many paraconsistent logics including da Costa's C-systems can be interpreted as the subclass of LFIs. Therefore, we can regard LFIs as a general framework for paraconsistent logics.

A Logic of Formal Inconsistency, which extends classical logic C with the consistency operator $\circ$, is defined as any explosive paraconsistent logic, namely iff the classical consequence relation $\vdash$ satisfies the following two conditions:

(a) $\exists\Gamma\exists A\exists B(\Gamma, A, \neg A \nvdash B)$
(b) $\forall\Gamma\forall A\forall B)(\Gamma, \circ A, A, \neg A \vdash B)$.

Here, Γ denotes a set of formulas and A, B are formulas. With the help of $\circ$, we can express both consistency and inconsistency in the object-language. Therefore, LFIs are general enough to classify many paraconsistent logics.

For example, da Costa's C_1 is shown to be an LFI. For every formula A, let $\circ A$ be an abbreviation of the formula $\neg(A \wedge \neg A)$. Then, the logic C_1 is an LFI such that $\circ(p) = \{\circ p\} = \{\neg\neg(p \wedge \neg p)\}$ whose axiomatization as an LFI contains the positive fragment of classical logic with the axiom $\neg\neg A \to A$, and some axioms for $\circ$.

(bc1) $\circ A \to (A \to (\neg A \to B))$
(ca1) $(\circ A \wedge \circ B) \to \circ(A \wedge B)$
(ca2) $(\circ A \wedge \circ B) \to \circ(A \vee B)$
(ca3) $(\circ A \wedge \circ B) \to \circ(A \to B)$

In addition, we can define classical negation $\sim$ by $\sim A =_{def} \neg A \wedge \circ A$. If needed, the inconsistency operator $\bullet$ is introduced by definition: $\bullet A =_{def} \neg \circ A$.

Carnielli et al. [27] showed classifications of existing logical systems. For example, classical logic is not an LFI, and Jáskowski's D_2 is an LFI. They also introduced a basic system of LFI, called LFI1 with a semantics and axiomatization.

We can thus see that the Logics of Formal Inconsistency are very interesting from a logical point of view in that they can serve as a theoretical framework for existing paraconsistent logics. In addition, there are tableau systems for LFIs; see

Fig. 2.3 The bilattice *FOUR*

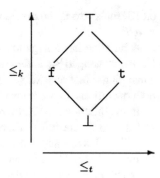

Carnielli and Marcos [25], and they can be properly applied to various areas including computer science and AI.

The above mentioned logics have been worked as a paraconsistent logic. But there are other logics which share the features of paraconsistent logics. The two notable examples are *possibilistic logic* and logics based on *bilattices*. In fact, these logics can properly deal both with incompleteness and inconsistency of information.

A *bilattice* was originally introduced by Ginsberg [35, 36] for the foundations of reasoning in AI, which has two kinds of orderings, i.e., truth ordering and knowledge ordering.

Later, it was extensively studied by Fitting in the context of logic programming in [33] and of theory of truth in [34]. In fact, bilattice-based logics can handle both incomplete and inconsistent information.

A *pre-bilattice* is a structure $\mathcal{B} = \langle B, \leq_t, \leq_k \rangle$, where B denotes a non-empty set and $\leq_t$ and $\leq_k$ are partial orderings on B. The ordering $\leq_k$ is thought of as ranking "degree of information (or knowledge)". The bottom in $\leq_k$ is denoted by $\bot$ and the top by $\top$. If $x <_k y$, y gives us at least as much information as x (and possibly more).

The ordering $\leq_t$ is an ordering on the "degree of truth". The bottom in $\leq_t$ is denoted by *false* and the top by *true*. A bilattice can be obtained by adding certain assumptions for connections for two orderings.

One of the most well-known bilattices is the bilattice *FOUR* as depicted as Fig. 2.3. The bilattice *FOUR* can be interpreted a combination of Belnap's lattices **A4** and **L4** as is clear from Fig. 2.3.

The bilattice *FOUR* can be seen as Belnap's lattice *FOUR* with two kinds of orderings. Thus, we can think of the left-right direction as characterizing the ordering $\leq_t$: a move to the right is an increase in truth.

The meet operation $\wedge$ for $\leq_t$ is then characterized by: $x \wedge y$ is rightmost thing that is of left both x and y. The join operation $\vee$ is dual to this. In a similar way, the up-down direction characterizes $\leq_k$: a move up is an increase in information. $x \otimes y$ is the uppermost thing below both x and y, and $\oplus$ is its dual.

Fitting [33] gave a semantics for logic programming using bilattices. Kifer and Subrahmanian [41] interpreted Fitting's semantics within generalized annotated logics *GAL*. Fitting [34] tried to generalize Kripke's [40] theory of truth, which is based on Kleene's strong three-valued logic, in a four-valued setting based on the bilattice *FOUR*.

A billatice has a negation operation $\neg$ if there is a mapping $\neg$ that reverse $\leq_t$, leaves unchanged $\leq_k$ and $\neg\neg x = x$. Likewise a bilattice has a *conflation* if there is a mapping—that reverse $\leq_k$, leaves unchanged $\leq_t$. and $--x = x$. If a bilattice has both operations, they *commute* if $--\neg x = \neg - x$ for all x.

In the billatice *FOUR*, there is a negation operator under which $\neg t = f$, $\neg f = t$, and $\bot$ and $\top$ are left unchanged. There is also a conflation under which $-\bot = \top$, $-\top = \bot$ and t and f are left unchanged. And negation and conflation commute. In any bilattice, if a negation or conflation exists then the extreme elements $\bot, \top, f$ and t will behave as in *FOUR*.

Bilattice logics are theoretically elegant in that we can obtain several algebraic constructions, and are also suitable for reasoning about incomplete and inconsistent information. Arieli and Avron [13, 14] studied reasoning with bilattices. Thus, bilattice logics have many applications in AI as well as philosophy.

Annotated logic is a logic for paraconsistent logic programming; see Subrahmanian [24, 52]. It is also regarded as one of the attractive paraconsistent logics; see da Costa et al [30, 31]. Note that annotated logic has many applications for several areas including engineering. And Abe studied annotated logic for many years.

Billatice logics described above are seen as a rival to annotated logics. We can also unify annotated logics and billatice logics; see Rico [50]. We will review annotated logic in details in Chap. 5; see Abe, Akama and Nakamatsu [1, 2].

Finally, we make an important remark. The propositional calculus is the basis of the usual classical and non-classical logics; however, a true and strong logical system has to contain quantification and a theory of identity at least, and should in principle incorporate a higher-order logic (a form of higher-order logic, some set theory or some other more or less equivalent logical tool).

The relevance of people like Frege, Russell and Peirce, is that they created quantification theory and other aspects of logic beyond the propositional level. Da Costa was the first logician to present a system of paraconsistent logic in this extended sense.

Acknowledgments The authors would like to thank the referee for constructive remarks.

References

1. Abe, J.M.: On the Foundations of Annotated Logics (in Portuguese), Ph.D. Thesis, University of São Paulo, Brazil (1992)
2. Abe, J.M., Akama, S., Nakamatsu, K.: Introduction to Annotated Logics. Springer, Heidelberg (2016)
3. Akama, S.: Resolution in constructivism. Logique et Analyse **120**, 385–399 (1987)
4. Akama, S.: Constructive predicate logic with strong negation and model theory. Notre Dame J. Formal Logic **29**, 18–27 (1988)
5. Akama, S.: On the proof method for constructive falsity. Zeitschrift für mathematische Logik und Grundlagen der Mathematik **34**, 385–392 (1988)

6. Akama, S.: Subformula semantics for strong negation systems. J. Philos. Logic **19**, 217–226 (1990)
7. Akama, S.: Constructive Falsity: Foundations and Their Applications to Computer Science, Ph.D. Thesis, Keio University, Yokohama, Japan (1990)
8. Akama, S.: Nelson's paraconsistent logics. Logic Logical Philos. **7**, 101–115 (1999)
9. Akama, S., Abe, J.M., Nakamatsu, K.: Constructive discursive logic with strong negation. Logique et Analyse **215**, 395–408 (2011)
10. Almukdad, A., Nelson, D.: Constructible falsity and inexact predicates. J. Symbolic Logic **49**, 231–233 (1984)
11. Anderson, A., Belnap, N.: Entailment: The Logic of Relevance and Necessity I. Princeton University Press, Princeton (1976)
12. Anderson, A., Belnap, N., Dunn, J.: Entailment: The Logic of Relevance and Necessity II. Princeton University Press, Princeton (1992)
13. Arieli, O., Avron, A.: Reasoning with logical bilattices. J. Logic Lang. Inform. **5**, 25–63 (1996)
14. Arieli, O., Avron, A.: The value of fur values. Artif. Intell. **102**, 97–141 (1998)
15. Arruda, A.I.: A survey of paraconsistent logic. In: Arruda, A., da Costa, N., Chuaqui, R. (eds.) Mathematical Logic in Latin America, pp. 1–41. North-Holland, Amsterdam (1980)
16. Asenjo, F.G.: A calculus of antinomies. Notre Dame J. Formal Logic **7**, 103–105 (1966)
17. Batens, D.: Dynamic dialectical logics. In: Priest, G., Routley, R. Norman, J. (eds.) Paraconsistent Logic: Essay on the Inconsistent, pp. 187–217. Philosophia Verlag, München (1989)
18. Batens, D.: Inconsistency-adaptive logics and the foundation of non-monotonic logics. Logique et Analyse **145**, 57–94 (1994)
19. Batens, D.: A general characterization of adaptive logics. Logique et Analyse **173–175**, 45–68 (2001)
20. Batens, D., Mortensen, C., Priest, G., Van Bendegem, J.-P. (eds.): Frontiers of Paraconsistent Logic. Research Studies Press, Baldock (2000)
21. Belnap, N.D.: A useful four-valued logic. In: Dunn, J.M., Epstein, G. (eds.) Modern Uses of Multi-Valued Logic, pp. 8–37. Reidel, Dordrecht (1977)
22. Belnap, N.D.: How a computer should think. In: Ryle, G. (ed.) Contemporary Aspects of Philosophy, pp. 30–55. Oriel Press (1977)
23. Beziau, J.-Y., Carnielli, W., Gabbay, D. (eds.): Handbook of Paraconsistency. College Publication, London (2007)
24. Blair, H.A., Subrahmanian, V.S.: Paraconsistent logic programming. Theoret. Comput. Sci. **68**, 135–154 (1989)
25. Carnielli, W.A., Marcos, J.: Tableau systems for logics of formal inconsistency. In: Abrabnia, H.R. (ed.), Proceedings of the 2001 International Conference on Artificial Intelligence, vol. II, pp. 848–852. CSREA Press (2001)
26. Carnielli, W.A., Coniglio, M.E., D'Ottaviano, I.M. (eds.): Paraconsistency: The Logical Way to the Inconsistent. Marcel Dekker, New York (2002)
27. Carnielli, W.A., Coniglio, M.E, Marcos, J.: Logics of formal inconsistency. In: Gabbay, D., Guenthner, F. (eds.) Handbook of Philosophical Logic, vol. 14, 2nd edn, pp. 1–93. Springer, Heidelberg (2007)
28. da Costa, N.C.A.: On the theory of inconsistent formal systems. Notre Dame J. Formal Logic **15**, 497–510 (1974)
29. da Costa, N.C.A., Alves, E.H.: A semantical analysis of the calculi C_n. Notre Dame J. Formal Logic **18**, 621–630 (1977)
30. da Costa, N.C.A., Abe, J.M., Subrahmanian, V.S.: Remarks on annotated logic. Zeitschrift für mathematische Logik und Grundlagen der Mathematik **37**, 561–570 (1991)
31. da Costa, N.C.A., Subrahmanian, V.S., Vago, C.: The paraconsistent logic $P\mathcal{T}$. Zeitschrift für mathematische Logik und Grundlagen der Mathematik **37**, 139–148 (1991)
32. Dunn, J.M.: Relevance logic and entailment. In: Gabbay, D., Gunthner, F. (eds.) Handbook of Philosophical Logic, vol. III, pp. 117–224. Reidel, Dordrecht (1986)
33. Fitting, M.: Bilattices and the semantics of logic programming. J. Logic Program. **11**, 91–116 (1991)

34. Fitting, M.: A theory of truth that prefers falsehood. J. Philos. Logic **26**, 477–500 (1997)
35. Ginsberg, M.: Multivalued logics. In: Proceedings of AAAI'86, pp. 243–247. Morgan Kaufman, Los Altos (1986)
36. Ginsberg, M.: Multivalued logics: a uniform approach to reasoning in AI. Comput. Intell. **4**, 256–316 (1988)
37. Jaśkowski, S.: Propositional calculus for contradictory deductive systems (in Polish). Studia Societatis Scientiarun Torunesis, Sectio A **1**, 55–77 (1948)
38. Jaśkowski, S.: On the discursive conjunction in the propositional calculus for inconsistent deductive systems (in Polish). Studia Societatis Scientiarun Torunesis, Sectio A **8**, 171–172 (1949)
39. Kleene, S.: Introduction to Metamathematics. North-Holland, Amsterdam (1952)
40. Kripke, S.: Outline of a theory of truth. J. Philos. **72**, 690–716 (1975)
41. Kifer, M., Subrahmanian, V.S.: On the expressive power of annotated logic programs. In: Proceedings of the 1989 North American Conference on Logic Programming, pp. 1069–1089 (1989)
42. Kotas, J.: The axiomatization of S. Jaskowski's discursive logic. Studia Logica **33**, 195–200 (1974)
43. Łukasiewicz, J.: On 3-valued logic. In: McCall, S. (ed.) Polish Logic, pp. 16–18, Oxford University Press, Oxford, 1967
44. Nelson, D.: Constructible falsity. J. Symbolic Logic **14**, 16–26 (1949)
45. Nelson, D.: Negation and separation of concepts in constructive systems. In: Heyting, A. (ed.) Constructivity in Mathematics, pp. 208–225. North-Holland, Amsterdam (1959)
46. Priest, G.: Logic of paradox. J. Philos. Logic **8**, 219–241 (1979)
47. Priest, G.: Paraconsistent logic. In: Gabbay, D., Guenthner, F. (eds.) Handbook of Philosophical Logic, 2nd edn, pp. 287–393. Kluwer, Dordrecht (2002)
48. Priest, G.: In Contradiction: A Study of the Transconsistent, 2nd edn. Oxford University Press, Oxford (2006)
49. Priest, G., Routley, R., Norman, J. (eds.): Paraconsistent Logic: Essays on the Inconsistent. Philosophia Verlag, München (1989)
50. Rico, G.O.: The annotated logics OP_{BL}. In: Carnielli, W., Coniglio, M., D'Ottaviano, I. (eds.) Paraconsistency: The Logical Way to the Inconsistent, pp. 411–433. Marcel Dekker, New York (2002)
51. Routley, R., Plumwood, V., Meyer, R.K., Brady, R: Relevant Logics and Their Rivals, vol. 1. Ridgeview, Atascadero (1982)
52. Subrahmanian, V.: On the semantics of quantitative logic programs. In: Proceeding of the 4th IEEE Symposium on Logic Programming, pp. 173–182 (1987)
53. van Fraassen, B.C.: Facts and tautological entailment. J. Philos. **66**, 477–487 (1069)
54. Vasil'ev, N.A.: Imaginary Logic (in Russian). Nauka, Moscow (1989)
55. Wansing, H.: The Logic of Information Structures. Springer, Berlin (1993)

Chapter 3
An Application of Paraconsistent Logic to Physics: Complementarity

Newton C.A. da Costa and Décio Krause

> *The apparently incompatible sorts of information about the behavior of the object under examination which we get by different experimental arrangements can clearly not be brought into connection with each other in the usual way, but may, as equally essential for an exhaustive account of all experience, be regarded as 'complementary' to each other.*
> Niels Bohr (1937), p. 291

Abstract In this paper we review some of the main ideas of two previous papers of ours which deal with an application of a kind of paraconsistent logic to quantum physics [14, 15]. We think that this revision is justified to present once more the richness of paraconsistent logics and suggests a way of dealing with one of most intriguing concepts of quantum theory. We propose and interpretation of complementarity in terms of what we call C-theories (theories involving the idea of complementarity, in a sense explained in the text), whose underlying logic is a kind of paraconsistent logic termed *paraclassical logic*. Roughly speaking, C-theories which may have 'physically' incompatible theorems (and, in particular, contradictory theorems), but which are not trivial.

Keywords Complementarity · Paraconsistency · Paraclassical logic

3.1 Introduction

The concept of 'complementarity' was introduced in quantum mechanics by Niels Bohr in his famous 'Como Lecture', in 1927 [2]. This concept is one of the core concepts of the Copenhagen interpretation of quantum mechanics [1, 22].

N.C.A. da Costa (✉) · D. Krause
Department of Philosophy, Federal University of Santa Catarina, Florianópolis,
SC, Brazil
e-mail: ncacosta@terra.com.br

D. Krause
e-mail: krause.decio@ufsc.br

© Springer International Publishing Switzerland 2016
S. Akama (ed.), *Towards Paraconsistent Engineering*, Intelligent Systems
Reference Library 110, DOI 10.1007/978-3-319-40418-9_3

Notwithstanding their importance, Bohr's ideas on complementarity are controversial, and there is no general agreement on a precise meaning of the *Principle of Complementarity*. As Bohr has advanced, "I think that it would be reasonable to say that no man who is called a philosopher really understands what is meant by complementary descriptions" (quoted from [11, p. 32]). This suggests the difficulties involved in any attempt elucidate the meaning of this Principle. Anyhow, this remark invites us to look also at the logico-mathematical grounds, mainly in connection with the paraconsistent program [17].

In this paper, we continue the investigation of a kind of theory which we termed C-theories [14, 15]. This is a way to find an application of paraconsistent logics to physics, although the concept here presented can be applied to other fields as well. In the mentioned papers, we have enlighten the main motivations taken from Bohr himself so as from some people close related to him who tried to explain the notion of complementarity in some way, including logic.

Let us begin with the concept of a C-theory, where the main motivation is revised.

3.2 C-theories

Let us quote Max Jammer in order to achieve a way to understand the notion of complementarity (in our two papers, we have quoted Bohr himself in several passages we shall not repeat here):

> Although it is not easy, as we see, to define Bohr's notion of *complementarity*, the notion of *complementarity interpretation* seems to raise fewer definitory difficulties. The following definition of this notion suggests itself. A given theory T admits a complementarity interpretation if the following conditions are satisfied: (1) T contains (at least) two descriptions D_1 and D_2 of its substance-matter; (2) D_1 and D_2 refer to the same universe of discourse U (in Bohr's case, microphysics); (3) neither D_1 nor D_2, if taken alone, accounts exhaustively for all phenomena of U; (4) D_1 and D_2 are mutually exclusive in the sense that their combination into a single description would lead to logical contradictions.

> That these conditions characterize a complementarity interpretation as understood by the Copenhagen school can easily be documented. According to Léon Rosenfeld, (...) one of the principal spokesmen of this school, complementarity is the answer to the following question: What are we to do when we are confronted with such situation, in which we have to use two concepts that are mutually exclusive, and yet both of them necessary for a complete description of the phenomena? "Complementarity denotes the logical relation, of quite a new type, between concepts which are mutually exclusive, and which therefore cannot be considered at the same time –that would lead to logical mistakes—but which nevertheless must both be used in order to give a complete description of the situation." Or to quote Bohr himself concerning condition (4): "In quantum physics evidence about atomic objects by different experimental arrangements (...) appears contradictory when combination into a single picture is attempted." (...) In fact, Bohr's Como lecture with its emphasis on the mutual exclusive but simultaneous necessity of the causal (D_1) and the space-time description (D_2), that is, Bohr's first pronouncement of his complementarity interpretation, forms an example which fully conforms with the preceding definition. Borh's discovery of complementarity, it is often said, constitutes his greatest contribution to the philosophy of modern science. [22, pp. 104–5]

We interpret Jammer's quotation as follows. Firstly, we shall take for granted that both D_1 and D_2 are sentences formulated in the language of a theory T and that they refer to the same universe of discourse, so that D_1 and D_2 can be formulated in its language. So, items (1) and (2) will be considered only implicitly. Item (3) will be understood as entailing that *both* D_1 and D_2 are, from the point of view of T, *necessary* for the full comprehension of the relevant aspects of the objects of the domain; so, we shall take both D_1 and D_2 as 'true' sentences (in an adequate 'model' of T). Concerning item (4), we recall that Jammer says that 'mutually exclusive' means that the "combination of D_1 and D_2 into a single description would lead to logical contradictions", and this is reinforced by Rosenfeld's words that the concepts "cannot be considered at the same time", since this would entail a "logical mistake". Then, we will informally say that *mutually exclusive*, or *complementary*, are incompatible sentences or propositions whose conjunction lead to a contradiction (in a theory T based on classical logic).

So, we say that a theory T is a C-theory, if T encompasses non equivalent theses α and β (which may stand for Jammer's D_1 and D_2 respectively) such that their conjunction yields to a contradiction in T, according to classical logic. As we shall see, it is not necessary that α and β be the negation of one another. In fact, the most interesting cases are precisely not this one, but those where α and β *entail* each of them propositions that are contradictory. Below we present a rigorous formulation of this idea.

The problem with the above characterization of complementary sentences is that if the underlying logic of T is classical logic or any other 'standard' logical system like intuitionistic logic, then T becomes contradictory, or inconsistent. Apparently, it is precisely this what Rosenfeld claimed in the above quotation. Obviously, if we intend to maintain the idea of complementary propositions in the sense just described, so we change the underlying logic of T by modifying the notion of deduction. Our motivation is a previous paper of one of us [16].

3.3   The Logic of C-theories

We will describe here just the propositional logic of C-theories, leaving the problem of extending it to quantification to be further investigated.

Let C be an axiomatized system of the classical propositional calculus [24]. The concept of deduction of C is the standard one; we use the symbol $\vdash$ to represent deductions in C. The formulas of C are denoted by Greek lowercase letters, while Greek uppercase letters stand for sets of formulas. The symbols $\neg$, $\rightarrow$, $\wedge$, $\vee$ and $\leftrightarrow$ have their usual meanings, and standard conventions in the writing of formulas will be also assumed without further comments. All the syntactical concepts and details are the standard ones [24]. In particular, we are interested in the following definitions:

Definition 3.1 A theory T (a set of formulas closed under deduction) is **inconsistent** if it contains a theorem α whose negation $\neg\alpha$ is also a theorem of T; otherwise, T is **consistent**.

Definition 3.2 If $\mathcal{F}$ denotes the set of all formulas of the language of C, then T is **trivial** if the set of its theorems coincides with $\mathcal{F}$; otherwise, T is **nontrivial**.

All syntactical concepts of P are similar to the corresponding concepts of C. The notion of *paraclassical deduction* is introduced as follows:

Definition 3.3 Let Γ be a set of formulas of P and let α be a formula (of the language of P). Then we say that α is a (syntactical) P-consequence of Γ, and write

$$\Gamma \vdash_\mathsf{P} \alpha$$

if and only if

(P1) $\alpha \in \Gamma$, or
(P2) There exists a consistent (according to classical logic) subset $\Delta \subseteq \Gamma$ such that
       $\Delta \vdash \alpha$ (in classical logic).

We call $\vdash_\mathsf{P}$ the relation of **P-consequence**. It is immediate that, among others, the following results can be proved:

Theorem 3.1 *1. If α is a theorem of the classical propositional calculus C and if Γ is a set of formulas, then $\Gamma \vdash_\mathsf{P} \alpha$. In particular, $\vdash_\mathsf{P} \alpha$.*
2. If Γ is consistent (according to C), then $\Gamma \vdash \alpha$ (in C) iff $\Gamma \vdash_\mathsf{P} \alpha$ (in P).
3. If $\Gamma \vdash_\mathsf{P} \alpha$ and if $\Gamma \subseteq \Delta$, then $\Delta \vdash_\mathsf{P} \alpha$ (The defined notion of P-consequence is monotonic).
4. The notion of P-consequence ($\vdash_\mathsf{P}$) is recursive.
5. Since the theses of P are the theses of C, P is decidable.

Definition 3.4 A set of formulas Γ is **P-trivial** iff $\Gamma \vdash_\mathsf{P} \alpha$ for every formula α. Otherwise, Γ is **P-nontrivial**.

Theorem 3.2 *Other results are the following ones, which reinforce the nature of our system:*

1. For any formula α and any set of formulas Γ, we have that $\Gamma \nvdash_\mathsf{P} \alpha \wedge \neg\alpha$.
2. $\alpha \vdash_\mathsf{P} \alpha$ iff α has a model.
3. If $\mathcal{F}$ denotes the set of all formulas, then $\mathcal{F} \nvdash_\mathsf{P} \alpha \wedge \neg\alpha$ for any formula α.
4. $\Gamma \vdash_\mathsf{P} \alpha$ and $\Gamma \vdash_\mathsf{P} \neg\alpha$ don't imply $\Gamma \vdash_\mathsf{P} \alpha \wedge \neg\alpha$.
5. Let p be a propositional variable. Then:

   (a) $p, \neg p \vdash_\mathsf{P} p$
   (b) $p, \neg p \vdash_\mathsf{P} \neg p$
   (c) $p, \neg p \nvdash_\mathsf{P} p \wedge \neg p$
   (d) $p \wedge \neg p \nvdash_\mathsf{P} p \wedge \neg p$

(e) $p \wedge \neg p \nvdash_P p$
(f) $p \wedge \neg p \nvdash_P \neg p$

Definition 3.5 A set of formulas Γ is **P-inconsistent** if there exists a formula α such that $\Gamma \vdash_P \alpha$ and $\Gamma \vdash_P \neg\alpha$. Otherwise, Γ is **P-consistent**.

Theorem 3.3 *1. If α is an atomic formula, then $\Gamma = \{\alpha, \neg\alpha\}$ is P-inconsistent, but P-nontrivial.*
2. If the set of formulas Γ is P-trivial, then it is trivial (according to classical logic). If Γ is nontrivial, then it is P-nontrivial.
3. If Γ is P-inconsistent, then it is inconsistent according to classical logic, but not conversely.
4. If Γ is consistent according to classical logic, then Γ is P-consistent.
5. The sole P-consequences of $p \wedge \neg p$ are tautologies (of classical logic).
6. The Deduction Theorem holds for P-consequences: $\Gamma, \alpha \vdash_P \beta$ entails $\Gamma \vdash_P \alpha \rightarrow \beta$.

We remark that $\{\alpha \wedge \neg\alpha\}$ is trivial in classical logic, but not P-trivial. Notwithstanding, we are not suggesting that complementary propositions should be understood as pairs of contradictory sentences.

Definition 3.6 A $\mathcal{C}$-**theory** is a set of formulas T closed under the relation of P-consequence $\vdash_P$, that is, $\alpha \in T$ for whatever α such that $T \vdash_P \alpha$. In other words, T is a theory whose underlying logic is P.

Theorem 3.4 *There exist C-theories that are inconsistent from the point of view of classical logic, though P-nontrivial.*

Proof Immediate consequence of Theorem 3.3. ⊣
In the common applications, the existence of consistent sets of formulas are usually assumed only in an informal way, as an implicit postulate. Intuitively speaking, it makes reference to the fact that some 'classical' (that is, based on usual mathematics) theories and hypotheses scientists accept are thought of as not contradictory (as consistent) in principle.

Theorem 3.5 *Every consistent classical theory, that is, every consistent theory founded in classical logic (and set theory) is a particular case of C-theories.*

Finally, we state a result (the theorem below), whose proof is an immediate consequence of the above definition of P-consequence, that links our logic with the characterization of 'complementary propositions' presented above. Before this, we make a definition:

Definition 3.7 Let T be a $\mathcal{C}$-theory and let α and β be formulas of the language of T. We say that α and β are T-**complementary** (or simply **complementary**) if there exists a formula γ of the language of T such that:

1. $T \vdash_P \alpha$ and $T \vdash_P \beta$
2. $\alpha \vdash_P \gamma$ and $\beta \vdash_P \neg \gamma$

It is immediate that contradictory propositions like α and $\neg \alpha$ are complementary in the above sense, but once more we remark that we are not arguing that this particular logical situation constitute a condensed account of all Bohr's ideas. The interesting case results from the following theorem.

Theorem 3.6 *If α and β are complementary theorems of a C-theory T and $\alpha \vdash_P \gamma$ and $\beta \vdash_P \neg \gamma$, then in general $\gamma \wedge \neg \gamma$ is not a theorem of T.*

Proof Immediate, as a consequence of Theorem 3.3.   ⊣

This result is in fact interesting, since we may admit propositions (complementary propositions) so that one of them entails a proposition while the another one entails the negation of such a proposition, but we cannot deduce that their conjunction entails a contradiction.

Now we sketch in a not so precise terms an example of a situation involving C-theories, suppose that our theory T is orthodox non-relativistic quantum mechanics, and that we have the following sentences, which can be written in the formalism:

1. (p) The system has a particle behavior.
2. (q) The system has a wave behavior.

For instance, p may be taken from a double slit experiment with one of the slits closed, while q stands for the experiment with the two slits open. It seems clear that we cannot have both p and q, that is, their conjunction. This is achieved in our system simply by verifying that, for instance, $p \vdash p$, that is, a particle behavior entails a particle behavior, while $q \vdash_P \neg p$, that is, a wave behavior excludes a particle behavior. The conjunction $p \wedge q$, then, within classical logic, would entail a contradiction. But not in our system, for there is not, in classical logic, a set of sentences Γ that entail propositions which, taken jointly, entail a contradiction. So, no set Δ of formulas can be found in order to satisfy Definition 3.7.

The basic characteristic of T as a C-theory is that in making inferences, we suppose that some hypotheses we handle are consistent. In other words, C-theories are closer to those theories scientists *actually* use in their day-to-day activity than theories encompassing the classical concept of deduction. We intend to discuss with more details in another work this and other related problems in the foundations of quantum mechanics, as for instance the reality of the wave function, to which paraclassical logic may be of some help.

3.4   The Paralogic Associated to a Logic $\mathcal{L}$

The technique used in this paper to define the paraclassical logic associated with classical logic can be generalized to any logic $\mathcal{L}$ (including logics having no negation symbol, but we will not deal with this case here). More precisely, starting from a logic $\mathcal{L}$, we can define the $\mathsf{P}_{\mathcal{L}}$-logic associated to $\mathcal{L}$ (the 'paralogic' associated to $\mathcal{L}$) as follows.

Let $\mathcal{L}$ be a logic, which may be classical logic, intuitionistic logic, some paraconsistent logic or, in principle, any other logical system containing negation. The deduction symbol of $\mathcal{L}$ is $\vdash_{\mathcal{L}}$, and it is defined according to the standards of the particular logic being considered. We still suppose that the language of $\mathcal{L}$ has a symbol for negation, $\neg$.

Definition 3.8 A **theory** based on $\mathcal{L}$ (an $\mathcal{L}$-**theory**) is a set of formulas Γ of the language of $\mathcal{L}$ which is closed under $\vdash_{\mathcal{L}}$. In other words, $\alpha \in \Gamma$ for every formula α such that $\Gamma \vdash_{\mathcal{L}} \alpha$.

Definition 3.9 An $\mathcal{L}$-theory Γ is $\mathcal{L}$-**inconsistent** if there exists a formula α of the language of $\mathcal{L}$ such that $\Gamma \vdash_{\mathcal{L}} \alpha$ and $\Gamma \vdash_{\mathcal{L}} \neg\alpha$, where $\neg\alpha$ is the negation of α. Otherwise, Γ is $\mathcal{L}$-**consistent**.

Definition 3.10 A $\mathcal{L}$-theory Γ is $\mathcal{L}$-**trivial** if $\Gamma \vdash_{\mathcal{L}} \alpha$ for any formula α of the language of $\mathcal{L}$. Otherwise, Γ is $\mathcal{L}$-**nontrivial**.

Then, we define the $\mathsf{P}_{\mathcal{L}}$-logic associated with $\mathcal{L}$ whose language and syntactical concepts are those of $\mathcal{L}$ but by modifying the concept of deduction as follows: we say that α is a $\mathsf{P}_{\mathcal{L}}$-**syntactical consequence** of a set Γ of formulas, and write $\Gamma \vdash_{\mathsf{P}_{\mathcal{L}}} \alpha$ iff:

1. $\alpha \in \Gamma$, or
2. There exists $\Delta \subseteq \Gamma$ such that Δ is $\mathcal{L}$-nontrivial, and $\Delta \vdash_{\mathcal{L}} \alpha$.

For instance, we may consider the paraconsistent calculus $\mathcal{C}_1$ [17] as our logic $\mathcal{L}$. Then the paralogic associated with $\mathcal{C}_1$ is a kind of 'para-paraconsistent' logic.

It seems worthwhile to note the following in connection with the paraclassical treatment of theories. Sometimes, when one has a paraclassical theory T such that $T \vdash_{\mathsf{P}} \alpha$ and $T \vdash_{\mathsf{P}} \neg\alpha$, there exist *appropriate* propositions β and γ such that T can be replaced by a classical consistent theory T' in which $\beta \rightarrow \alpha$ and $\gamma \rightarrow \neg\alpha$ are theorems. If this happens, the logical difficulty is in principle eliminable and classical logic maintained.

3.5   More General Complementary Situations

We conclude the same way as we did in one of our papers, for we think that what we have said there remains still valid in a general sense.

As it is well known, Bohr tried to apply his principle of complementarity to other fields of knowledge [22]. More recently, Englert et al. [19] have suggested that complementarity is not simply a consequence of the uncertainty relations, as advocated by those who believe that "two complementary variables, such as position and momentum, cannot simultaneously be measured to less than a fundamental limit of accuracy" (op. cit.), but that

(. . .) uncertainty is not the only enforce of complementarity. We devised and analysed both real and thought experiments that bypass the uncertainty relation, in effect to 'trick' the quantum objects under study. Nevertheless, the results always reveal that nature safeguards itself against such intrusions—complementarity remains intact even when the uncertainty relation plays no role. We conclude that complementarity is deeper than has been appreciated: it is more general and more fundamental to quantum mechanics than is the uncertainty rule. (ibid.)

If Englert et al. are right, then it seems that paraclassical logic can be useful also to treat those theories which encompass complementarity in their sense.

Anyway, this kind of logic can be also modified to cope with more general kinds of incompatibility, say 'physical incompatibility', incorporating physical incompatible postulates, so as characteristics of the behavior of human beings, etc., but we shall leave this topic for another work.

3.6   Final Remarks

The referee of this paper has asked us to answer some interesting questions, namely:

1. Can C-theories be applied to other problems, e.g., Schrödinger's cat and the EPR paradox?
2. Is there a proof theory for C-theories ? Is it possible by natural deduction?
3. Can Jaśkowski's discursive logic also be used for the basis for C-theories?

Of course the answers would demand another paper. But we would like to say yes to the last two questions, and leaving open the first. Let us give some brief explanations why.

Concerning the last two questions, it seems clear that a natural deduction system can also be developed, whose rules would be inspired in our axiomatics, and that a proof theory *should* be developed, so as to extend these logics to quantification and perhaps to higher-order logics. As for Jaśkowski's logic, since it is also non-adjunctive, it could of course be another alternative, although a very complicated one. Paraclassical logics are simple for the purposes of physics, and we guess they should be preferred.

Concerning the first question, it depends on what we understand by a contradiction in the quantum domain. The authors of this paper disagree in this respect. For da Costa and Christian de Ronde [13], Schrödinger's cat is a typical example of a contradiction in the quantum domain. For Krause and Arenhart [23], this is not so, for once we assume that we can speak of the cat before measurement (so leaving out a purely instrumentalistic interpretation such as Bohr's), when in the superposed state the cat is not alive *and* dead, contrary to the common understanding. To these authors, the cat is in a third state, precisely that state which is typical of quantum mechanics, the superposed state. So, there would be three possible states for the cat: dead cat, alive cat, and a superposed state of these two ones. In short, 'dead cat' is not the contradictory of 'alive cat', but it is its *contrary*, according to the square of oppositions. In the mentioned papers, a full discussion is provided defending each position. The problem concerning EPR apparently can be treated in a similar way, but a further analysis is recommended in this case.

Acknowledgments We would like to thank the organizers of this volume for the opportunity of presenting this paper and dedicate it to Jair Minoro Abe, our friend and colleague of so many years.

References

1. Beller, M.: The birth of Bohr's complementarity: the context and the dialogues. Stud. Hist. Phil. Sci. **23**(1), 147–180 (1992)
2. Bohr, N.: The quantum postulate and the recent development of atomic theory' [3], Atti del Congresso Internazionale dei Fisici, 11–20 Sept. 1927, Como-Pavia-Roma, Vol. II, Zanichelli, Bologna, 1928, reprinted in [pp.109–136]boh85
3. Bohr, N.: The quantum postulate and the recent developments of atomic theory. Nature **121**(Suppl.), 580–590 (1928), reprinted in [pp.147-158]boh85
4. Bohr, N.: 'Introductory survey' to Bohr (1929), in [pp. 279–302] boh85
5. Bohr, N.: *Atomic theory and the description of nature*. Cambridge University Press, Cambridge (1934). Reprinted in [pp. 279–302]boh85
6. Bohr, N.: Causality and complementarity. Phil. Sci. **4**(3), 289–298 (1937)
7. Bohr, N.: Natural philosophy of human cultures (1938). In: Atomic physics and human knowledge, pp. 23–31. Wiley, New York (1958) (also in Nature **143**, 268–272)
8. Bohr, N.: Quantum physics and philosophy: causality and complementarity. In: Klibanski, R. (ed.) Philosophy in the Mid-Century I, pp. 308–314. Firenze, La Nuova Italia (1958)
9. Bohr, N.: Collected works. In: Rüdinger, E. (general ed.), vol. 6: Foundations of Quantum Physics I. (1985) Kolckar, J. (ed.) Amsterdam, North-Holland
10. Carnap, R.: An Introduction to the Philosophy of Science. Dover Pu, New York (1995)
11. Cushing, J.T.: Quantum Mechanics: Historical Contingency and the Copenhagen Hegemony. The University of Chicago Press, Chicago and London (1994)
12. da Costa, N.C.A., Bueno, O.: Paraconsistency: towards a tentative interpretation. Theoria-Segunda Época **16**(1), 119–145 (2001)
13. da Costa, N.C.A., de Ronde, C.: The paraconsistent logic of superpositions. Found. Phys. **43**, 854–858 (2013)
14. da Costa, N.C.A., Krause, D.: Complementarity and paraconsistency. In: Rahman, S., Symons, J., Gabbay, D.M., van Bendegem, J.-P. (eds.) Logic, epistemology, and the Unity of Science, pp. 557–568. Springer (2004)
15. da Costa, N.C.A., Krause, D.: The logic of complementarity. In: van Benthem, J., Heinzmann, G., Rebushi, M., Visser, H. (eds.) The Age of Alternative Logics: Assessing Philosophy of Logic and Mathematics Today, pp. 103–120. Springer (2006)
16. da Costa, N.C.A., Vernengo, R.J.: Sobre algunas lógicas paraclássicas y el análisis del razonamiento jurídico. Doxa **19**, 183–200 (1999)
17. da Costa, N.C.A., Krause, D., Bueno, O.: Paraconsistent logics and paraconsistency. In: D. Jacquette, editor of the volume on Philosophy of Logic; Gabbay, D.M., Thagard, P., Woods, J. (eds.) Philosophy of Logic, Elsevier, 2006, in the series Handbook of the Philosophy of Science, vol. 5, pp. 655–781 (2007)
18. De Février, P.: La structure des théories physiques. Presses Un, de France, Paris (1951)
19. Englert, B.-G., Scully, M.O., Walther, H.: The duality in matter and light. Sci. Am. **271**(6), 56–61 (1994)
20. French, A.P., Kennedy, P.J. (eds.): Niels Bohr, a Centenary Volume. Harward University Press, Cambridge, MA and London (1985)

21. Hughes, G.E., Creswell, M.J.: A New Introduction to Modal Logic. Routledge, London (1996)
22. Jammer, M.: Philosophy of Quantum Mechanics. John Wiley, New York (1974)
23. Krause, D., Arenhart, J.R.B.: A logical account of quantum superpositions', forthtcoming. In: Aerts, D., de Ronde, C., Freytes, H., Giuntini, R. (eds.) Probing the Meaning and Structure of Quantum Mechanics: Superpositions, Semantics, Dynamics and Identity. World Scientific, Singapore (2016)
24. Mendelson, E.: Introduction to Mathematical Logic, 3rd edn. Wadsworth & Brooks/Cole, Monterrey (1987)
25. Scheibe, E.: The Logical Analisys of Quantum Mechanics. Pergamon Press, Oxford (1973)

Chapter 4
Two Genuine 3-Valued Paraconsistent Logics

Jean-Yves Beziau

Dedicated to Jair Minoro Abe for his 60th birthday

Abstract In this paper we present two genuine three-valued paraconsistent logics, i.e. logics obeying neither $p, \neg p \vdash q$ nor $\vdash \neg(p \wedge \neg p)$. We study their basic properties and their relations with other paraconsistent logics, in particular da Costa's paraconsistent logics $C1$ and its extension $C1+$.

Keywords Paraconsistent logic · Many-valued logic · Negation

4.1 Genuine Paraconsistent Negation

A paraconsistent negation is often defined on the basis of the rejection of the law of explosion. This means there are propositions p and q such that:

$$p, \neg p \nvdash q$$

As we have emphasized in previous papers (see [7, 8]), on the one hand such a negative definition of paraconsistent negation, taking alone, is nonsense, because there are many operators which are non explosive and cannot be reasonably considered

J.-Y. Beziau (✉)
UFRJ—Federal University of Rio de Janeiro, Rio de Janeiro, Brazil
e-mail: jyb@jyb-logic.org

J.-Y. Beziau
CNPq—Brazilian Research Council, Rio de Janeiro, Brazil

© Springer International Publishing Switzerland 2016
S. Akama (ed.), *Towards Paraconsistent Engineering*, Intelligent Systems
Reference Library 110, DOI 10.1007/978-3-319-40418-9_4

as negations, such as for example the operator of possibility. On the other hand this negative criterion of rejection of explosion is not enough. For example I. Urbas has pointed out that it makes sense also to reject a weakest form of explosion, such as:

$$p, \neg p \nvdash \neg q$$

This eliminates minimal negation from the realm of paraconsistent negations. Urbas gave a general formulation of the rejection of the explosion which leads to its definition of *strict paraconsistent logic*, his idea is that it is not possible to deduce any non tautological scheme of formula from p and $\neg p$ (see [24]).

It also makes sense to have the following rejection for a paraconsistent negation:

$$\nvdash \neg(p \wedge \neg p)$$

because it is a quite natural formulation of the principle of non contradiction. If one wants to study a negation rejecting the principle of non contradiction and develop a system of logic where $\neg(p \wedge \neg p)$ is a theorem, he has to explain what it means. "Philosophical Logic" is an expression used to qualify the study of systems of logic having a philosophical inspiration or/and motivation, but very often the connection between the formalism and its philosophical aspect is quite loose, if not incoherent (This is in particular the case in paraconsistent logic, see e.g. [10]).

In a recent published paper [11] we have called *strong* paraconsistent negation, a negation rejecting this version of the principle of non contradiction as well as explosion. We have replaced here *strong* by *genuine*, because A. Avron pointed to us that *strong* has already been used in another sense: as synonymous to *strict*. We have started to systematically study which kinds of genuine paraconsistent negations can be built using three-valued matrices. This is a new line of research because although some general frameworks have been presented for three-valued paraconsistent logics (cf. [2, 19]), the particular three-valued paraconsistent logics which have been studied up to now generally have $\neg(p \wedge \neg p)$ as a theorem. In the present paper we are further on developing this line of research, presenting new results, but the paper is self-contained.

4.2   Two Genuine Three-Valued Paraconsistent Logics

After a systematic study of all the possibilities (see [11]), we have distinguished two interesting genuine three-valued paraconsistent logics, which are defined by the following truth-tables:

SP3A

¬	
0	1
①	1
1	0

∧	0	①	1
0	0	0	0
①	0	①	1
1	0	1	1

∨	0	①	1
0	0	①	1
①	①	①	1
1	1	1	1

SP3B

¬	
0	1
①	①
1	0

∧	0	①	1
0	0	0	0
①	0	1	①
1	0	①	1

∨	0	①	1
0	0	①	1
①	①	①	1
1	1	1	1

We are using three values 0, 1 and ①. 1 and ① are both considered as designated and 0 as undesignated. Using these tables, we define in the usual way two logical structures $SP3A = \langle \mathbb{L}; \vdash_{SP3A} \rangle$ and $SP3B = \langle \mathbb{L}; \vdash_{SP3B} \rangle$. These two consequence relations are therefore structural consequence relations in the sense of Łoś and Suszko [17].

The reason we use ① as designated rather than $\frac{1}{2}$ is to let open the possible interpretation of this third value, which is not necessarily at the middle of the two other ones. However in the tables we had to choose the position and we have placed the third value at the middle. The truth-table of disjunction of both *SP3A* and *SP3B* follows the idea that the value of the disjunction of two propositions is the greatest value of the two propositions if we consider ① as in-between. But what is happening with the tables of conjunction of *SP3A* and *SP3B* is not the standard idea that the value of the conjunction of two propositions is the least value, if we consider ① as in-between. If we want to follow the idea of the least value, in the table for conjunction of *SP3A* ① should be rather seen as 2, but then this is not compatible with the interpretation of ① in the table for disjunction of *SP3A*. In the case of the table of conjunction of *SP3B*, there is no way to interpreted ① to follow the idea of the least value, because the value of the conjunction of two propositions having both ① as truth-value is 1.

In the history of three-valued logic, the third value has been philosophically interpreted in different ways, for example as *undetermined* or *possible* (cf. Łukasiewicz— see [18]) and generally 0 and 1 as respectively *falsity* and *truth*. However the distinction between designated and undesignated values keeps a dichotomy which is similar to the dichotomy between truth and falsity, this manifests in particular when this dichotomy is used to define logical truth and a consequence relation. So we can say that ① is a kind of truth or part of truth, this is why we have used this notation.[1]

---

[1]In [11] we have used *k*+ because we were also examining the possibilities where the third value was undesignated, using the notation *k*–.

The truth-tables for negation of *SP3A* and *SP3B* are different but in both cases *p* and ¬*p* can both be designated, both be true. This is in fact what always happens in paraconsistent logic following a standard semantical interpretation of the rejection of explosion. This is why a paraconsistent negation cannot be considered as forming contradiction, i.e. it does not make sense to say that *p* and ¬*p* is a contradiction in paraconsistent logic (see [3, 9]).

The three-valued paraconsistent logics developed by Asenjo [1] (renamed LP by Priest [20]) and da Costa—D'Ottaviano (the logic J3 [14]) have a paraconsistent negation defined by the the the same table as the one of *SP3B* and also the same table for disjunction, but they use for conjunction a table based on the least value. For this reason they are not genuine paraconsistent logics, because ¬$(p \land \neg p)$ is a theorem. These logics are using in fact the same basic truth-tables as "classical" three-valued logics developed by Łukasiewicz [18] or Kleene [16]. The only difference is that they consider the third value as designated. Sette's logic P1 [21] uses for negation the same truth-table as the one of *SP3A*, but he uses for conjunction and disjunction tables where the third value disappears, as a consequence any molecular formula behaves classically in this logic.

4.3 Basic Properties of *SP3A* and *SP3B*

4.3.1 Conjunction and Disjunction

The properties of conjunction and disjunction of *SP3A* and *SP3B* are the same as in classical logic despite the fact that the table of conjunction of *SP3B* is weird by itself and that the table of conjunction of *SP3A* is weird in relation with the table of disjunction. These tables have three important properties:

• they are conservative, i.e. the part concerning classical values 0 and 1 is the same as in classical logic;
• they are neo-classical, i.e. the dichotomy designated/undesignated behaves like the dichotomy 0/1 in classical logic, e.g. $\land(x, y)$ is designated iff *x* is designated and *y* is designated;
• they are symmetrical, i.e. $*(x, y) = *(y, x)$.

4.3.2 Laws of Negations that **SP3A** and **SP3B** Do Not Obey

Working in the framework of a structural consequence relation, if we reject the law of explosion, in its various formulations, some properties of negations are immediately not valid, for example:

Contraposition
   If $T, p \vdash q$ then $T, \neg q \vdash \neg p$
Reduction to the the Absurd
   If $T, \neg p \vdash q$ and $T, \neg p \vdash \neg q$ then $T \vdash p$
For a general study of the laws of negation and their interrelations, see [6]. On the other hand despite these negative results, many properties of classical negation are compatible with the rejection of the law of explosion, this is why it makes sense to still speak of a *negation* when dealing with paraconsistency. We will see in the next sections positive properties of the negations of *SP3A* and *SP3B*.

4.3.3 Excluded Middle

The law of excluded middle holds both for *SP3A* and *SP3B*, as shown in the following tables:

p	$\neg p$	$p \vee \neg p$
0	1	1
①	1	1
1	0	1

SP3A

p	$\neg p$	$p \vee \neg p$
0	1	1
①	①	①
1	0	1

SP3B

4.3.4 Double Negation

The double negation law $p \dashv\vdash \neg\neg p$ holds for *SP3B* but not for *SP3A*, as shown by the following tables:

p	$\neg p$	$\neg\neg p$
0	1	0
①	1	0
1	0	0

SP3A

p	$\neg p$	$\neg\neg p$
0	1	0
①	①	①
1	0	1

SP3B

In *SP3A*, given an atomic formula a, we have $a \nvdash \neg\neg a$.

4.3.5 De Morgan Laws

We will present De Morgan laws in three steps. Here is a first table:

D1a	$\neg(p \wedge q) \vdash \neg p \vee \neg q$	D1b	$\neg p \vee \neg q \vdash \neg(p \wedge q)$
D2a	$\neg(p \vee q) \vdash \neg p \wedge \neg q$	D2b	$\neg p \wedge \neg q \vdash \neg(p \vee q)$

De Morgan Laws—Table I

SP3A and *SP3B* obey all these laws except *D1b* as shown by the two following tables:

p	q	$\neg p$	$\neg q$	$p \wedge q$	$\neg(p \wedge q)$	$\neg p \vee \neg q$	$\neg p \wedge \neg q$	$\neg(p \vee q)$	$p \vee q$
0	①	1	1	0	1	1	1	1	①
①	0	1	1	0	1	1	1	1	①
①	①	1	1	①	1	1	1	1	①
①	1	1	0	1	0	1	0	0	1
1	①	0	1	1	0	1	0	0	1

De Morgan Laws—Table I—Checking the Validity for *SP3A*

p	q	$\neg p$	$\neg q$	$p \wedge q$	$\neg(p \wedge q)$	$\neg p \vee \neg q$	$\neg p \wedge \neg q$	$\neg(p \vee q)$	$p \vee q$
0	①	1	①	0	1	1	①	①	①
①	0	①	1	0	1	1	①	①	①
①	①	①	①	1	0	①	1	①	①
①	1	①	0	①	①	①	0	0	1
1	①	0	①	①	①	①	0	0	1

De Morgan Laws—Table I—Checking the Validity for *SP3B*

Here is a second table for De Morgan laws:

D3a	$\neg(\neg p \wedge q) \vdash p \vee \neg q$	D3b	$p \vee \neg q \vdash \neg(\neg p \wedge q)$
D4a	$\neg(p \wedge \neg q) \vdash \neg p \vee q$	D4b	$\neg p \vee q \vdash \neg(p \wedge \neg q)$
D5a	$\neg(\neg p \vee q) \vdash p \wedge \neg q$	D5b	$p \wedge \neg q \vdash \neg(\neg p \vee q)$
D6a	$\neg(p \vee \neg q) \vdash \neg p \wedge q$	D6b	$\neg p \wedge q \vdash \neg(p \vee \neg q)$

De Morgan Laws—Table II

Note that considering commutativity D3 and D4 are equivalent, same remark about D5 and D6.

p	q	$\neg p$	$\neg q$	$\neg p \wedge q$	$\neg(\neg p \wedge q)$	$p \vee \neg q$	$p \wedge \neg q$	$\neg(\neg p \vee q)$	$\neg p \vee q$
0	①	1	1	1	0	1	0	0	1
①	0	1	1	0	1	1	1	0	1
①	①	1	1	1	0	1	1	0	1
①	1	1	0	1	0	①	0	0	1
1	①	0	1	0	1	1	1	1	①

De Morgan Laws—Table II—Checking the Validity for *SP3A*

According to this table, D3a, D4a, D5a, D6a are valid for for *SP3A* but not D3b, D4b, D5b, D6b.

p	q	$\neg p$	$\neg q$	$\neg p \wedge q$	$\neg(\neg p \wedge q)$	$p \vee \neg q$	$p \wedge \neg q$	$\neg(\neg p \vee q)$	$\neg p \vee q$
0	Ⓘ	1	Ⓘ	Ⓘ	Ⓘ	Ⓘ	0	0	1
Ⓘ	0	Ⓘ	1	0	1	1	Ⓘ	Ⓘ	Ⓘ
Ⓘ	Ⓘ	Ⓘ	Ⓘ	1	0	Ⓘ	1	Ⓘ	Ⓘ
Ⓘ	1	Ⓘ	0	Ⓘ	Ⓘ	Ⓘ	0	0	1
1	Ⓘ	0	Ⓘ	0	1	1	Ⓘ	Ⓘ	Ⓘ

De Morgan Laws—Table II—Checking the Validity for *SP3B*

According to this table, D3a, D4a, D5a, D6a, D5b, D6b are valid for for *SP3B* but not D3b, D4b.

And here is a third table for De Morgan laws:

D7a	$\neg(\neg p \wedge \neg q) \vdash p \vee q$	D7b	$p \vee q \vdash \neg(\neg p \wedge \neg q)$
D8a	$\neg(\neg p \vee \neg q) \vdash p \wedge q$	D8b	$p \wedge q \vdash \neg(\neg p \vee \neg q)$

De Morgan Laws—Table III

p	q	$\neg p$	$\neg q$	$\neg p \wedge \neg q$	$\neg(\neg p \wedge \neg q)$	$p \vee q$	$p \wedge q$	$\neg(\neg p \vee \neg q)$	$\neg p \vee \neg q$
0	Ⓘ	1	1	1	0	Ⓘ	0	0	1
Ⓘ	0	1	1	1	0	Ⓘ	0	0	1
Ⓘ	Ⓘ	1	1	1	0	Ⓘ	Ⓘ	0	1
Ⓘ	1	1	0	0	1	1	1	0	1
1	Ⓘ	0	1	0	1	1	1	0	1

De Morgan Laws—Table III—Checking the Validity for *SP3A*

According to this table, D7a, D8a for for *SP3A* but not D7a, D8a.

p	q	$\neg p$	$\neg q$	$\neg p \wedge \neg q$	$\neg(\neg p \wedge \neg q)$	$p \vee q$	$p \wedge q$	$\neg(\neg p \vee \neg q)$	$\neg p \vee \neg q$
0	Ⓘ	1	Ⓘ	Ⓘ	Ⓘ	Ⓘ	0	0	1
Ⓘ	0	Ⓘ	1	Ⓘ	Ⓘ	Ⓘ	0	0	1
Ⓘ	Ⓘ	Ⓘ	Ⓘ	1	0	Ⓘ	1	Ⓘ	Ⓘ
Ⓘ	1	Ⓘ	0	0	1	1	Ⓘ	Ⓘ	Ⓘ
1	Ⓘ	0	Ⓘ	0	1	1	Ⓘ	Ⓘ	Ⓘ

De Morgan Laws—Table III—Checking the Validity for *SP3B*

According to this table, D7a, D8a for *SP3B* but not D7a, D8a. We see that *SP3A* and *SP3B* both verifies all the Da laws. They also both verifies D2b. *SP3B* additionaly verifies D5b and D6b and therefore is stronger than *SP3A* regarding De Morgan laws.

4.3.6 Definition of a Classical Negation

We say that a (scheme of) formula behaves classically when p is designated iff $\neg p$ is undesignated. In both $SP3A$ and $SP3B$ $p \wedge \neg p$ behaves classically. We say that a unary $\times$ connective behaves as a classical negation when p is designated iff $\times p$ is undesignated. In both $SP3A$ and $SP3A$ $\neg^* p = \neg p \wedge \neg(p \wedge \neg p)$ behaves as a classical negation. This is shown by the following tables:

p	$\neg p$	$p \wedge \neg p$	$\neg(p \wedge \neg p)$	$\neg p \wedge \neg(p \wedge \neg p)$
0	1	0	1	1
①	1	1	0	0
1	0	0	1	0

Classical negation in $SP3A$

p	$\neg p$	$p \wedge \neg p$	$\neg(p \wedge \neg p)$	$\neg p \wedge \neg(p \wedge \neg p)$
0	1	0	1	1
①	①	1	0	0
1	0	0	1	0

Classical negation in $SP3B$

As a consequence it is possible to translate classical logic into $SP3A$ and $SP3B$. These are more examples of the translation paradox [15]. The classical negation defined here is the same as the one of da Costa's system $C1$ (see [13]). In the next section we will study the relations between $SP3A$, $SP3B$ and $C1$.

4.4 Comparison with da Costa Paraconsistent Logics $C1$ and $C1+$

The paraconsistent logic $C1$ introduced by Newton da Costa (see [12]) can be constructed on the basis of positive classical logic adding the following laws for negation (see [5]):

$\vdash p \vee \neg p$

$p \wedge \neg p, \neg(p \wedge \neg p) \vdash q$

$\neg\neg p \vdash p$

$p^\circ, q^\circ \vdash (p©q)^\circ$

where © is any binary connective[2] and p° is an abbreviation for $\neg(p \wedge \neg p)$.
We have proposed a strengthening $C1+$ of the logic $C1$ by considering instead of the last axioms, the following one (see [4]):

$x^\circ \vdash (p©q)^\circ$, x being either p or q.

---

[2]© can be implication. In the present paper we are studying logics without implication, so we are comparing these logics with the fragment of $C1$ without implication.

The idea of the original axiom of $C1$ is that if two formulas are well-behaved, then it is also the case of compositions of them using binary connectives. In $C1+$ the idea is that only one formula needs to be well-behaved to have the well-behavior of a complex formula. We use the terminology *weak preservation* for the axiom of $C1$ and *strong preservation* for the axiom of $C1+$. In $C1+$ we have consequently more De Morgan laws which are valid.

Here is a comparative table (all converses are not valid):

C1	C1+
$\neg(p \wedge q) \vdash \neg p \vee \neg q$	$\neg(p \vee q) \vdash \neg p \wedge \neg q$
$\neg(\neg p \wedge \neg q) \vdash p \vee q$	$\neg(\neg p \vee \neg q) \vdash p \wedge q$
$\neg(p \wedge \neg q) \vdash \neg p \vee q$	$\neg(p \vee \neg q) \vdash \neg p \wedge q$
$\neg(\neg p \wedge q) \vdash p \vee \neg q$	$\neg(\neg p \vee q) \vdash p \wedge \neg q$

Considering the results of the previous section on De Morgan laws for $SP3A$ and $SP3B$, we can draw the following comparative tables:

$\neg(p \wedge q) \vdash \neg p \vee \neg q$	$C1+$ $SP3A$ $SP3B$ $C1$
$\neg(\neg p \wedge \neg q) \vdash p \vee q$	$C1+$ $SP3A$ $SP3B$ $C1$
$\neg(p \wedge \neg q) \vdash \neg p \vee q$	$C1+$ $SP3A$ $SP3B$ $C1$
$\neg(\neg p \wedge q) \vdash p \vee \neg q$	$C1+$ $SP3A$ $SP3B$ $C1$
$\neg(p \vee q) \vdash \neg p \wedge \neg q$	$C1+$ $SP3A$ $SP3B$
$\neg(\neg p \vee \neg q) \vdash p \wedge q$	$C1+$ $SP3A$ $SP3B$
$\neg(p \vee \neg q) \vdash \neg p \wedge q$	$C1+$ $SP3A$ $SP3B$
$\neg(\neg p \vee q) \vdash p \wedge \neg q$	$C1+$ $SP3A$ $SP3B$

Comparative table I of De Morgan Laws
in $C1+$ $SP3A$ $SP3B$ $C1$

$\neg p \vee \neg q \vdash \neg(p \wedge q)$	None
$p \vee q \vdash \neg(\neg p \wedge \neg q)$	None
$\neg p \vee q \vdash \neg(p \wedge \neg q)$	None
$p \vee \neg q \vdash \neg(\neg p \wedge q)$	None
$\neg p \wedge \neg q \vdash \neg(p \vee q)$	$SP3A$ $SP3B$
$p \wedge q \vdash \neg(\neg p \vee \neg q)$	None
$\neg p \wedge q \vdash \neg(p \vee \neg q)$	$SP3B$
$p \wedge \neg q \vdash \neg(\neg p \vee q)$	$SP3B$

Comparative table II of De Morgan Laws
in $C1+$ $SP3A$ $SP3B$ $C1$

Relatively to De Morgan Laws we have therefore the following relation of order between these four logics:

$C1 < C1+ < SP3A < SP3B$

Except the preservation axiom, we have already seen that all the axioms of $C1$ are valid in $SPA3$ and $SP3B$. Let us examine now both the weak and strong preservation axioms for *SPA*3 and *SP*3B. To simplify the notation, we introduce $p^•$ for the formula $p \wedge \neg p$. Therefore we have $p^° = \neg p^•$

In *SP*3A, we have the following table:

p	q	$\neg p$	$p^•$	$p^°$	$p \wedge q$	$\neg(p \wedge q)$	$(p \wedge q)^•$	$(p \wedge q)^°$
0	①	1	0	1	0	1	0	1
①	0	1	1	0	0	1	0	1
①	①	1	1	0	①	1	1	0
①	1	1	1	0	1	0	0	1
1	①	0	0	1	1	0	0	1

Table for strong preservation of
well-behavior under conjunction in *SP*3A

This table shows that the strong axiom for preservation under conjunction $x^° \vdash (p \wedge q)^°$ is valid in *SP*3A, therefore the weak axiom $p^°, q^° \vdash (p \wedge q)^°$ is also valid. In *SP*3B, we have the following table:

p	q	$\neg p$	$p^•$	$p^°$	$p \wedge q$	$\neg(p \wedge q)$	$(p \wedge q)^•$	$(p \wedge q)^°$
0	①	1	0	1	0	1	0	1
①	0	①	1	0	0	1	0	1
①	①	①	1	0	1	0	0	1
①	1	①	1	0	①	①	1	0
1	①	0	0	1	①	①	1	0

Table for strong preservation of
well-behavior under conjunction in *SP*3B

The fourth line of truth values of this table shows that the strong axiom for preservation under conjunction $x^° \vdash (p \wedge q)^°$ is not valid in *SP*3B. The following table shows however that the weak axiom is valid:

p	q	$\neg p$	$\neg q$	$p^•$	$p^°$	$q^•$	$q^°$	$p \wedge q$	$\neg(p \wedge q)$	$(p \wedge q)^•$	$(p \wedge q)^°$
0	①	1	①	0	1	1	0	0	1	0	1
①	0	①	1	1	0	0	1	0	1	0	1
①	①	①	①	1	0	1	0	1	0	0	1
①	1	①	0	1	0	0	1	①	①	1	0
1	①	0	①	0	1	1	0	①	①	1	0

Table for weak preservation of well-behavior under conjunction in *SP*3B

This table shows that in *SP*3B the formulas $p^°$ and $q^°$ cannot be designated together (considering the non classical lines of the table), this means that the weak axiom for preservation under disjunction $p^°, q^° \vdash (p \vee q)^°$ is also valid in *SP*3B. The following table shows that the weak axiom for preservation under disjunction is also valid in *SP*3A for the same reason. It shows also that the strong axiom for preservation under disjunction $x^° \vdash (p \vee q)^°$ is not valid in *SP*3A as indicated by the first and the second lines of truth values.

p	q	$\neg p$	$\neg q$	$p^\bullet$	p°	$q^\bullet$	q°	$p \vee q$	$\neg(p \vee q)$	$(p \vee q)^\bullet$	$(p \vee q)^\circ$
0	①	1	1	0	1	1	0	①	1	1	0
①	0	1	1	1	0	0	1	①	1	1	0
①	①	1	1	1	0	1	0	①	1	1	0
①	1	1	0	1	0	0	1	1	0	0	1
1	①	0	1	0	1	1	0	1	0	0	1

Table for weak preservation of well-behavior under disjunction in *SP3A*

The following table shows that the strong strong axiom for preservation under disjunction $x^\circ \vdash (p \vee q)^\circ$ is not valid in *SP3B* as indicated by the first line of truth values..

p	q	$\neg p$	$p^\bullet$	p°	$p \vee q$	$\neg(p \vee q)$	$(p \vee q)^\bullet$	$(p \vee q)^\circ$
0	①	1	0	1	①	①	1	0
①	0	①	1	0	①	①	1	0
①	①	①	1	0	①	①	1	0
①	1	①	1	0	1	0	0	1
1	①	0	0	1	1	0	0	1

Table for strong preservation of well-behavior under disjunction in *SP3B*

We now summarize all the results in the following table:

$p^\circ, q^\circ \vdash (p \wedge q)^\circ$	C1+ SP3A SP3B C1
$p^\circ, q^\circ \vdash (p \vee q)^\circ$	C1+ SP3A SP3B C1
$x^\circ \vdash (p \wedge q)^\circ$	C1+ SP3A
$x^\circ \vdash (p \vee q)^\circ$	C1+

Comparative table of preservation of
well-behavior in *C1+ SP3A SP3B C1*

The conclusion is that *SP3A* and *SP3B* are extensions of *C1*, but thought *C1+* appears as a strict extension of *SP3A* and *SP3B* from the point of view of the preservation axioms, *C1+* is weaker from the point of view of De Morgan Laws, so *C1+* is incomparable. *SP3B* is stronger than *SP3A* from the point of view of De Morgan Laws but is weaker from the point of view of the preservation axioms, so *SP3A* and *SP3B* are incomparable.

4.4.1 Replacement Theorem

The replacement theorem says that if two propositions p and q are logically equivalent (i.e. $p \dashv\vdash q$) one can be replaced by the other one. Considering that both *SP3A* and *SP3B* are extensions of *C1*, we can infer that the replacement theorem does not hold for them, due to Urbas's theorem according to which there are no self-extensional paraconsistent extensions of *C1*, see [22, 23].

A concrete case of failure of replacement theorem (given to us by A.Avron) working both for *SP3A* and *SP3B* is the following: p is logically equivalent to $(p \vee \neg p) \wedge p$ but $\neg p$ is not logically equivalent to $\neg((p \vee \neg p) \wedge p)$.

This can be seen by the following table, the first line of truth-values corresponding to the situation in *SP3A* and the second to the situation in *SP3B*.

p	$\neg p$	$p \vee \neg p$	$(p \vee \neg p) \wedge p$	$\neg((p \vee \neg p) \wedge p)$
①	1	1	1	0
①	①	①	1	0

4.5 Comparison Table Between *SP3A* and *SP3B*

To finish we present in the following table a comparative study of the properties of *SP3A* and *SP3B*:

	SP3A	SP3B
$p, \neg p \vdash q$	No	No
$p, \neg p \vdash \neg q$	No	No
$\vdash \neg(p \wedge \neg p)$	No	No
$\vdash p \vee \neg p$	Yes	Yes
$\neg\neg p \vdash p$	Yes	Yes
$p \vdash \neg\neg p$	No	Yes
$\neg p, \neg\neg p \vdash q$	Yes	No
$p \wedge \neg p, \neg(p \wedge \neg p) \vdash q$	Yes	Yes
$p^\circ, q^\circ \vdash (p \wedge q)^\circ$	Yes	Yes
$p^\circ, q^\circ \vdash (p \vee q)^\circ$	Yes	Yes
$x^\circ \vdash (p \wedge q)^\circ$	Yes	No
$x^\circ \vdash (p \vee q)^\circ$	No	No
$\neg(p \wedge q) \vdash \neg p \vee \neg q$	Yes	Yes
$\neg(\neg p \wedge \neg q) \vdash p \vee q$	Yes	Yes
$\neg(p \wedge \neg q) \vdash \neg p \vee q$	Yes	Yes
$\neg(\neg p \wedge q) \vdash p \vee \neg q$	Yes	Yes
$\neg(p \vee q) \vdash \neg p \wedge \neg q$	Yes	Yes
$\neg(\neg p \vee \neg q) \vdash p \wedge q$	Yes	Yes
$\neg(p \vee \neg q) \vdash \neg p \wedge q$	Yes	Yes
$\neg(\neg p \wedge q) \vdash p \vee \neg q$	Yes	Yes
$\neg p \vee \neg q \vdash \neg(p \wedge q)$	No	No
$p \vee q \vdash \neg(\neg p \wedge \neg q)$	No	No
$\neg p \vee q \vdash \neg(p \wedge \neg q)$	No	No
$p \vee \neg q \vdash \neg(\neg p \wedge q)$	No	No
$\neg p \wedge \neg q \vdash \neg(p \vee q)$	Yes	Yes
$p \wedge q \vdash \neg(\neg p \vee \neg q)$	No	No
$\neg p \wedge q \vdash \neg(p \vee \neg q)$	No	Yes
$p \vee \neg q \vdash \neg(\neg p \vee q)$	No	Yes

Acknowledgments This paper was written during a stay at University of Tel Aviv within the GeTFun exchange prorgram—Marie Curie project PIRSES-GA-2012-318986 funded by EU-FP7. Thanks to Arnon Avron for his useful comments.

References

1. Asenjo, F.G.: A calculus of antinomies. Notre Dame J. Formal Logic **7**, 103–105 (1966)
2. Arieli, O., Avron, A.: Three-valued paraconsistent propositional logics. In: Beziau, J.-Y., Chakraborty, M., Dutta, S. (eds.) New Directions in Paraconsistent Logic, pp. 91–129. Springer, New Delhi (2015)
3. Becker, J.R.: Arenhart Liberating paraconsistency from contradictions. Log. Univers. **9**, 523–545 (2015)
4. Beziau, J.-Y.: Logiques construites suivant les méthodes de da Costa. Logique et Analyse **131–132**, 259–272 (1990)
5. Beziau, J.-Y.: Nouveaux résultats et nouveau regard sur la logique paraconsistante C1. Logique et Analyse **141–142**, 45–48 (1993)
6. Beziau, J.-Y.: Théorie législative de la négation pure. Logique et Analyse **147–148**, 209–225 (1994)
7. Beziau, J.-Y.: What is paraconsistent logic?. In: Batens, D. et al. (eds.) Frontiers of Paraconsistent Logic, pp. 95–111. Research Studies Press, Baldock (2000)
8. Beziau, J.-Y.: Are paraconsistent negations negations?. In: Carnielli, W. et al. (eds.) Paraconsistency: The Logical Way to the Inconsistent, pp. 465–486. Marcel Dekker, New-York (2002)
9. Beziau, J.-Y.: Round squares are no contractions. In: Beziau, J.-Y., Chakraborty, M., Dutta, S. (eds.) New Directions in Paraconsistent Logic, pp. 39–55. Springer, New Delhi (2015)
10. Beziau, J.-Y.: Trivial dialetheism and the logic of paradox. Logic Logical Philos. **25**, 51–56 (2016)
11. Beziau, J.-Y., Franschetto, A.: Strong paraconsistent three-valued logic. In: Beziau, J.-Y., Chakraborty, M., Dutta, S. (eds.) New Directions in Paraconsistent Logic, pp. 131–147. Springer, New Delhi (2015)
12. da Costa, N.C.A.: Calculs propositionnels pour les systèmes formels inconsistants. Cr. R. Acad Sc. Paris **257**, 3790–3793 (1963)
13. da Costa, N.C.A., Guillaume, M.: Négations composées et Loi de Peirce dans les systèmes Cn. Portugalia Mathematica **24**, 201–210 (1965)
14. D'Ottaviano, I.M.L., da Costa, N.C.A.: Sur un problème de Jaskowśki. Cr. R. Acad Sc. Paris **270**, 1349–1353 (1970)
15. Humberstone, L.: Beziau's translation paradox. Theoria **71**, 138–181 (2005)
16. Kleene, S.: On a notation for ordinal numbers. J. Symbolic. Logic **3**, 150–155 (1938)
17. Łoś, J., Suszko, R.: Remarks on sentential logics. Indigationes Mathematicae **10**, 177–183 (1958)
18. Łukasiewicz, J.: O logice trójwartościowej. Ruch Filozoficny **5**, 170–171 (1920)
19. Marcos, J.: 8K solutions and semi-solutions to a problem of da Costa. Unpublished manuscript (2000)
20. Priest, G.: The logic of paradox. J. Philos. Logic **8**, 219–241 (1979)
21. Sette, A.M.: On the propositional calculus P1. Notas e comunicacões de matemática, **17**, Recife (1971)
22. Urbas, I.: On Brazilian paraconsistent logics. Ph.D. Australian National University, Canberra (1987)
23. Urbas, I.: Paraconsistency and the C-systems of da Costa. Notre Dame J. Formal Logic **30**, 583–597 (1989)
24. Urbas, I.: Paraconsistency. Studies in Soviet Thought **39**, 343–354 (1989)

Chapter 5
A Survey of Annotated Logics

Seiki Akama

Dedicated to Jair Minoro Abe for his 60th birthday

Abstract Annotated logics have been originally developed as foundations for para-consistent logic programming, and later developed as paracomplete and paraconsistent logics by J.M. Abe and others. In this paper, we present the formalization of propositional and predicate annotated logics. We also review some formal issues.

Keywords Paraconsistent logics · Annotated logics · Paraconsistency · Paracompleteness · Paraconsistent logic programming

5.1 Introduction

One of J.M. Abe's contributions to paraconsistent logics is to establish the so-called *annotated logics*, which are paraconsistent and in general paracomplete. They have been developed as theoretical foundations for paraconsistent logic programming for inconsistent knowledge; see Subrahmanian [45] and Blair and Subrahmanian [22]. Later, they have been studied as the systems of paraconsistent logic by many people; see [1, 26, 30].

Abe explored many applications of annotated logics to various areas, including engineering. It is thus interesting to sketch the basics of annotated logics. We show their formal aspects without proofs. The complete exposion of annotated logics can be found in Abe et al. [8].

S. Akama (✉)
C-Republic, 1-20-1 Higashi-Yurigaoka, Asao-ku, Kawasaki 215-0012, Japan
e-mail: akama@jcom.home.ne.jp

© Springer International Publishing Switzerland 2016
S. Akama (ed.), *Towards Paraconsistent Engineering*, Intelligent Systems
Reference Library 110, DOI 10.1007/978-3-319-40418-9_5

The chapter is structured as follows. In Sect. 5.2, we present propositional annotated logics $P\tau$. In Sect. 5.3, we describe predicate annotated logics $Q\tau$. Section 5.4 gives Curry algebras as an algebraic semantics for annotated logics. We give some conclusions in Sect. 5.5.

5.2 Propositional Annotated Logics $P\tau$

As reviewed in Chap. 2, paraconsistent logics have been developed as the basis to formalize inconsistent but non-trivial theories, and many systems of paraconsistent logic have been proposed in the literature. Recently, we can find several interesting applications of paraconsistent logics for various areas including mathematics, philosophy and computer science.

There are historically three important systems of paraconsistent logic; see Priest et al. [40]. Jaśkowski proposed a paraconsistent propositional logic called *discursive logic* (or discussive logic) in 1948; see Jaśkowski [33, 34]. Discursive logic is a kind of modal approach to paraconsistency.

Da Costa proposed the so-called *C-system*, which is based on the non-standard interpretation of negation which is dual to intuitionistic negation. He developed propositional and predicate logic for C-system.

Relevance logic (or relevant logic) due to Anderson and Belnap formalizes a correct interpretation of implication, and some of relevant systems can be viewed as paraconsistent; see Anderson and Belnap [15] and Anderson et al. [16] and Routley et al. [44]. For a comprehensive survey, consult Dunn [31].

Since then, a lot of work has been done to develop a paraconsistent logic from some motivation. For a recent survey of paraconsistent logic, see Priest [42].

In 1979, Priest [41] proposed a *logic of paradox*, denoted LP, to deal with the semantic paradox.

Batens developed the so-called *adaptative logics* in Batens [18, 19] as improvements of *dynamic dialectical logics* developed in Batens [17]. *Inconsistency-adaptive logics* as developed by Batens [18] can be regarded as paraconsistent and nonmonotonic logics.

Carnelli's *Logics of Formal Inconsistency* (LFI) are logical systems that deal with consistency and inconsistency as object-level concept; see Carnelli et al. [23]. And several paraconsistent systems can be interpreted in LFI.

Now, we turn to a formal presentation of annotated logics. Before doing it, we introduce some basic concepts. Let T be a theory whose underlying logic is L. T is called *inconsistent* when it contains theorems of the form A and $\neg A$ (the negation of A), i.e.,

$$T \vdash_L A \text{ and } T \vdash_L \neg A$$

where $\vdash_L$ denotes the provability relation in L. If T is not inconsistent, it is called *consistent*.

T is said to be *trivial*, if all formulas of the language are also theorems of T. Otherwise, T is called *non-trivial*. Then, for trivial theory T, $T \vdash_L B$ for any formula B. Note that trivial theory is not interesting since every formula is provable.

If L is classical logic (or one of several others, such as intuitionistic logic), the notions of inconsistency and triviality agree in the sense that T is inconsistent iff T is trivial. So, in trivial theories the extensions of the concepts of formula and theorem coincide.

A *paraconsistent logic* is a logic that can be used as the basis for inconsistent but non-trivial theories. In this regard, paraconsistent theories do not satisfy, in general, the *principle of non-contradiction*, i.e., $\neg(A \wedge \neg A)$.

We can also define a paracomplete logic. A *paracomplete logic* is a logic, in which the *principle of excluded middle*, i.e., $A \vee \neg A$ is not a theorem. In this sense, intuitionistic logic is one of the paracomplete logics. A *paracomplete theory* is a theory based on paracomplete logic.

Finally, the logic which is simultaneously paraconsistent and paracomplete is called *non-alethic logic*.

The important problems handled by paraconsistent logics include the paradoxes of set theory, the semantic paradoxes, and some issues in dialectics. These problems are central to philosophy and philosophical logic. However, paraconsistent logics have later found interesting applications in AI, in particular, expert systems, belief, and knowledge, among others, since the 1980s; see da Costa and Subrahmanian [29].

Annotated logics were introduced by Subrahmanian to provide a foundation for paraconsistent logic programming; see Subrahmanian [45] and Blair and Subrahmanian [22]. Paraconsistent logic programming can be seen as an extension of logic programming based on classical logic.

In 1989, Kifer and Lozinskii proposed a logic for reasoning with inconsistency, which is related to annotated logics; see Kifer and Lozinskii [35, 36]. In the same year, Kifer and Subrahmanian extended annotated logics by introducing *generalized annotated logics* in the context of logic programming; see Kifer and Subrahmanian [37]. In 1990, a resolution-style automatic theorem-proving method for annotated logics was implemented; see da Costa et al. [28].

Of course, annotated logics were developed as a foundation for paraconsistent logic programming, but they have interesting features to be examined by logicians. Formally, annotated logics are ingeneral non-alethic in the sense of the above terminology. From a viewpoint of paraconsistent logicians, annotated logics were regarded as new systems.

In 1991, da Costa and others started to study annotated logics from a foundational point of view; see da Costa et al. [26, 30]. In these works, propositional and predicate annotated logics were formally investigated by presenting axiomatization, semantics and completeness results, and some applications of annotated logics were briefly surveyed.

In 1992, Jair Minoro Abe wrote Ph.D. thesis on the foundations of annotated logics under Prof. Newton C.A. da Costa at University of São Paulo; see Abe [1]. Abe proposed annotated modal logics which extend annotated logics with modal operator in Abe [2]; also see Akama and Abe [9].

Some formal results including decidability annotated logics were presented in
Abe and Akama [6]. Abe and Akama also investigated predicate annotated logics
by the method of ultraproducts in Abe and Akama [5]. Abe [3] studied an algebraic
semantics of annotated logics.

Now, we formally introduce annotated logics. The language of the propositional
annotated logics $P\tau$. We denote by L the language of $P\tau$. Annotated logics are
based on some arbitrary fixed finite lattice called a *lattice of truth-values* denoted by
$\tau = \langle |\tau|, \leq, \sim \rangle$, which is the complete lattice with the ordering $\leq$ and the operator
$\sim : |\tau| \rightarrow |\tau|$.

Here, $\sim$ gives the "meaning" of atomic-level negation of $P\tau$. We also assume that
$\top$ is the top element and $\bot$ is the bottom element, respectively. In addition, we use
two lattice-theoretic operations: $\vee$ for the least upper bound and $\wedge$ for the greatest
lower bound.[1]

Definition 5.1 (*Symbols*) The symbols of $P\tau$ are defined as follows:

1. Propositional symbols: $p, q, \ldots$ (possibly with subscript)
2. Annotated constants: $\mu, \lambda, \ldots \in |\tau|$
3. Logical connectives: $\wedge$ (conjunction), $\vee$ (disjunction), $\rightarrow$ (implication), and
   $\neg$ (negation)
4. Parentheses: (and)

Definition 5.2 (*Formulas*) Formulas are defined as follows:

1. If p is a propositional symbol and $\mu \in |\tau|$ is an annotated constant, then p_μ is a
   formula called an *annotated atom*.
2. If F is a formula, then $\neg F$ is a formula.
3. If F and G are formulas, then $F \wedge G$, $F \vee G$, $F \rightarrow G$ are formulas.
4. If p is a propositional symbol and $\mu \in |\tau|$ is an annotated constant, then a formula
   of the form $\neg^k p_\mu$ ($k \geq 0$) is called a *hyper-literal*. A formula which is not a hyper-
   literal is called a *complex formula*.

Here, some remarks are in order. The annotation is attached only at the atomic level.
An annotated atom of the form p_μ can be read "it is believed that p's truth-value is
at least μ". In this sense, annotated logics incorporate the feature of many-valued
logics.

A hyper-literal is special kind of formula in annotated logics. In the hyper-literal
of the form $\neg^k p_\mu$, $\neg^k$ denotes the k's repetition of $\neg$. More formally, if A is an
annotated atom, then $\neg^0 A$ is A, $\neg^1 A$ is $\neg A$, and $\neg^k A$ is $\neg(\neg^{k-1} A)$. The convention
is also use for $\sim$.

Next, we define some abbreviations.

---

[1] We employ the same symbols for lattice-theoretical operations as the corresponding logical connectives.

Definition 5.3 Let A and B be formulas. Then, we put:

$$A \leftrightarrow B =_{def} (A \to B) \land (B \to A)$$
$$\neg_* A =_{def} A \to (A \to A) \land \neg(A \to A)$$

Here, $\leftrightarrow$ is called the *equivalence* and $\neg_*$ *strong negation*, respectively.

Observe that strong negation in annotated logics behaves classically in that it has all the properties of classical negation.

We turn to a semantics for $P\tau$. We here describe a *model-theoretic semantics* for $P\tau$. Let $\mathbf{P}$ is the set of propositional variables. An *interpretation I* is a function $I : \mathbf{P} \to \tau$. To each interpretation I, we associate a *valuation* $v_I : \mathbf{F} \to \mathbf{2}$, where $\mathbf{F}$ is a set of all formulas and $\mathbf{2} = \{0, 1\}$ is the set of truth-values. Henceforth, the subscript is suppressed when the context is clear.

Definition 5.4 (*Valuation*) A valuation v is defined as follows:
If p_λ is an annotated atom, then

$v(p_\lambda) = 1$ iff $I(p) \geq \lambda$,
$v(p_\lambda) = 0$ otherwise,
$v(\neg^k p_\lambda) = v(\neg^{k-1} p_{\sim\lambda})$, where $k \geq 1$.

If A and B are formulas, then

$v(A \land B) = 1$ iff $v(A) = v(B) = 1$,
$v(A \lor B) = 0$ iff $v(A) = v(B) = 0$,
$v(A \to B) = 0$ iff $v(A) = 1$ and $v(B) = 0$.

If A is a complex formula, then

$v(\neg A) = 1 - v(A)$.

Say that the valuation v *satisfies* the formula A if $v(A) = 1$ and that v *falsifies* A if $v(A) = 0$. For the valuation v, we can obtain the following lemmas.

Lemma 5.1 *Let p be a propositional variable and $\mu \in |\tau|$ ($k \geq 0$), then we have:*

$$v(\neg^k p_\mu) = v(p_{\sim^k \mu}).$$

Lemma 5.2 *Let p be a propositional variable, then we have:*

$$v(p_\perp) = 1$$

Lemma 5.3 *For any complex formula A and B and any formula F, the valuation v satisfies the following:*

1. $v(A \leftrightarrow B) = 1$ iff $v(A) = v(B)$
2. $v((A \to A) \land \neg(A \to A)) = 0$
3. $v(\neg_* A) = 1 - v(A)$
4. $v(\neg F \leftrightarrow \neg_* F) = 1$

We here define the notion of semantic consequence relation denoted by $\models$. Let Γ be a set of formulas and F be a formula. Then, F is a *semantic consequence* of Γ, written $\Gamma \models F$, iff for every v such that $v(A) = 1$ for each $A \in \Gamma$, it is the case that $v(F) = 1$. If $v(A) = 1$ for each $A \in \Gamma$, then v is called a *model* of Γ. If Γ is empty, then $\Gamma \models F$ is simply written as $\models F$ to mean that F is *valid*.

Lemma 5.4 *Let p be a propositional variable and $\mu, \lambda \in |\tau|$. Then, we have:*

1. $\models p_\perp$
2. $\models p_\mu \to p_\lambda$, $\mu \geq \lambda$
3. $\models \neg^k p_\mu \leftrightarrow p_{\sim^k \mu}$, $k \geq 0$

The consequence relation $\models$ satisfies the next property.

Lemma 5.5 *Let A, B be formulas. Then, if $\models A$ and $\models A \to B$ then $\models B$.*

Lemma 5.6 *Let F be a formula and p a propositional variable. $(\mu_i)_{i \in J}$ be an annotated constant, where J is an indexed set. Then, if $\models F \to p_\mu$, then $F \to p_{\mu_i}$, where $\mu = \bigvee \mu_i$.*

As a corollary to Lemma 5.6, we can obtain the following lemma.

Lemma 5.7 $\models p_{\lambda_1} \wedge p_{\lambda_2} \wedge \cdots \wedge p_{\lambda_m} \to p_\lambda$, *where* $\lambda = \bigvee_{i=1}^{m} \lambda_i$.

Next, we discuss some results related to paraconsistency and paracompleteness.

Definition 5.5 (*Complementary property*) A truth-value $\mu \in \tau$ has the *complementary property* if there is a λ such that $\lambda \leq \mu$ and $\sim \lambda \leq \mu$. A set $\tau' \subseteq \tau$ has the *complementary property* iff there is some $\mu \in \tau'$ such that μ has the complementary property.

Definition 5.6 (*Range*) Suppose I is an interpretation of the language L. The *range* of I, denoted $range(I)$, is defined to be $range(I) = \{\mu \mid (\exists A \in B_L) I(A) = \mu\}$, where B_L denotes the set of all ground atoms in L.

For $P\tau$, ground atoms correspond to propositional variables. If the range of the interpretation I satisfies the complementary property, then the following theorem can be established.

Theorem 5.1 *Let I be an interpretation such that $range(I)$ has the complementary property. Then, there is a propositional variable p and $\mu \in |\tau|$ such that*

$$v(p_\mu) = v(\neg p_\mu) = 1.$$

Theorem 5.1 states that there is a case in which for some propositional variable it is both true and false, i.e., inconsistent. The fact is closely tied with the notion of paraconsistency.

Definition 5.7 (¬-*inconsistency*) We say that an interpretation I is ¬-*inconsistent* iff there is a propositional variable p and an annotated constant $\mu \in |\tau|$ such that $v(p_\mu) = v(\neg p_\mu) = 1$.

Therefore, ¬-inconsistency means that both A and $\neg A$ are simultaneously true for some atomic A. Below, we formally define the concepts of non-triviality, paraconsistency and paracompleteness.

Definition 5.8 (*Non-triviality*) We say that an interpretation I is *non-trivial* iff there is a propositional variable p and an annotated constant $\mu \in |\tau|$ such that $v(p_\mu) = 0$.

By Definition 5.8, we mean that not every atom is valid if an interpretation is non-trivial.

Definition 5.9 (*Paraconsistency*) We say that a interpretation I is *paraconsistent* iff it is both ¬-inconsistent and non-trivial. $P\tau$ is called *paraconsistent* iff there is an interpretation of I of $P\tau$ such that I is paraconsistent.

Definition 5.9 allows the case in which both A an $\neg A$ are true, but some formula B is false in some paraconsistent interpretation I.

Definition 5.10 (*Paracompleteness*) We say that an interpretation I is *paracomplete* iff there is a propositional variable p and a annotated constant $\lambda \in |\tau|$ such that $v(p_\lambda) = v(\neg p_\lambda) = 0$. $P\tau$ is called *paracomplete* iff there is an interpretation I of $P\tau$ such that I is paracomplete.

From Definition 5.10, we can see that in the paracomplete interpretation I, both A and $\neg A$ are false. We say that $P\tau$ is *non-alethic* iff it is both paraconsistent and paracomplete. Intuitively speaking, paraconsistent logic can deal with inconsistent information and paracomplete logic can handle incomplete information.

This means that non-alethic logics like annotated logics can serve as logics for expressing both inconsistent and incomplete information. This is one of the starting points of our study of annotated logics.

As the following Theorems 5.2 and 5.3 indicate, paraconsistency and paracompleteness in $P\tau$ depend on the cardinality of τ.

Theorem 5.2 *$P\tau$ is paraconsistent iff $card(\tau) \geq 2$, where $card(\tau)$ denotes the cardinality (cardinal number) of the set τ.*

Theorem 5.3 *If $card(\tau) \geq 2$, then there are annotated systems $P\tau$ such that they are paracomplete.*

The above two theorems imply that to formalize a non-alethic logic based on annotated logics we need at least both the top and bottom elements of truth-values. The simplest lattice of truth-values is $FOUR$ in Belnap [20, 21].

Definition 5.11 (*Theory*) Given an interpretation I, we can define the theory $Th(I)$ associated with I to be a set:

$$Th(I) = Cn(\{p_\mu \mid p \in \mathbf{P} \text{ and } I(p) \geq \mu\}).$$

Here, Cn is the semantic consequence relation, i.e.,

$$Cn(\Gamma) = \{F \mid F \in \mathbf{F} \text{ and } \Gamma \models F\}.$$

Here, Γ is a set of formulas.

$Th(I)$ can be extended for any set of formulas.

Theorem 5.4 *An interpretation I is $\neg$-inconsistent iff $Th(\Gamma)$ is $\neg$-inconsistent.*

Theorem 5.5 *An interpretation I is paraconsistent iff $Th(I)$ is paraconsistent.*

The next lemma states that the replacement of equivalent formulas within the scope of $\neg$ does not hold in $P\tau$ as in other paraconsistent logics.

Lemma 5.8 *Let A be any hyper-literal. Then, we have:*

1. $\models A \leftrightarrow ((A \rightarrow A) \rightarrow A)$
2. $\not\models \neg A \leftrightarrow \neg(((A \rightarrow A) \rightarrow A))$
3. $\models A \leftrightarrow (A \wedge A)$
4. $\not\models \neg A \leftrightarrow \neg(A \wedge A)$
5. $\models A \leftrightarrow (A \vee A)$
6. $\not\models \neg A \leftrightarrow \neg(A \vee A)$

As obvious from the above proofs, (1), (3) and (5) hold for any formula A. But, (2), (4) and (6) cannot be generalized for any A.

By the next theorem, we can find the connection of $P\tau$ and the positive fragment of classical propositional logic C.

Theorem 5.6 *If $F_1, \ldots, F_n$ are complex formulas and $K(A_1, \ldots, A_n)$ is a tautology of C, where $A_1, \ldots, A_n$ are the sole propositional variable occurring in the tautology, then $K(F_1, \ldots, F_n)$ is valid in $P\tau$. Here, $K(F_1, \ldots, F_n)$ is obtained by replacing each occurrence of A_i, $1 \leq i \leq n$, in K by F_i.*

Next, we consider the properties of strong negation $\neg_*$.

Theorem 5.7 *Let A, B be any formulas. Then,*

1. $\models (A \rightarrow B) \rightarrow ((A \rightarrow \neg_* B) \rightarrow \neg_* A)$
2. $\models A \rightarrow (\neg_* A \rightarrow B)$
3. $\models A \vee \neg_* A$

Theorem 5.7 tells us that strong negation has all the basic properties of classical negation. Namely, (1) is a principle of *reductio ad abusurdum*, (2) is the related principle of the law of non-contradiction, and (3) is the law of excluded middle. Note that $\neg$ does not satisfy these properties. It is also noticed that for any complex formula $A \models \neg A \leftrightarrow \neg_* A$ but that for any hyper-literal $Q \not\models \neg Q \leftrightarrow \neg_* Q$.

From these observations, $P\tau$ is a paraconsistent and paracomplete logic, but adding strong negation enables us to perform classical reasoning.

Next, we provide an axiomatization of $P\tau$ in the Hilbert style. There are many ways to axiomatize a logical system, one of which is the *Hilbert system*. Hilbert system can be defined by the set of *axioms* and *rules of inference*. Here, an axiom is a formula to be postulated as valid, and rules of inference specify how to prove a formula.

We are now ready to give a Hilbert style axiomatization of $P\tau$, called $\mathcal{A}\tau$. Let A, B, C be arbitrary formulas, F, G be complex formulas, p be a propositional variable, and λ, μ, λ_i be annotated constant. Then, the postulates are as follows (cf. Abe [1]):

Postulates for $\mathcal{A}\tau$

$(\rightarrow_1)\ (A \rightarrow (B \rightarrow A)$
$(\rightarrow_2)\ (A \rightarrow (B \rightarrow C)) \rightarrow ((A \rightarrow B) \rightarrow (A \rightarrow C))$
$(\rightarrow_3)\ ((A \rightarrow B) \rightarrow A) \rightarrow A$
$(\rightarrow_4)\ A, A \rightarrow B / B$
$(\wedge_1)\ (A \wedge B) \rightarrow A$
$(\wedge_2)\ (A \wedge B) \rightarrow B$
$(\wedge_3)\ A \rightarrow (B \rightarrow (A \wedge B))$
$(\vee_1)\ A \rightarrow (A \vee B)$
$(\vee_2)\ B \rightarrow (A \vee B)$
$(\vee_3)\ (A \rightarrow C) \rightarrow ((B \rightarrow C) \rightarrow ((A \vee B) \rightarrow C))$
$(\neg_1)\ (F \rightarrow G) \rightarrow ((F \rightarrow \neg G) \rightarrow \neg F)$
$(\neg_2)\ F \rightarrow (\neg F \rightarrow A)$
$(\neg_3)\ F \vee \neg F$
$(\tau_1)\ p_\perp$
$(\tau_2)\ \neg^k p_\lambda \leftrightarrow \neg^{k-1} p_{\sim\lambda}$
$(\tau_3)\ p_\lambda \rightarrow p_\mu$, where $\lambda \geq \mu$

$(\tau_4)\ p_{\lambda_1} \wedge p_{\lambda_2} \wedge \cdots \wedge p_{\lambda_m} \rightarrow p_\lambda$, where $\lambda = \bigvee_{i=1}^{m} \lambda_i$

Here, except $(\rightarrow_4)$, these postulates are axioms. $(\rightarrow_4)$ is a rule of inferences called *modus ponens* (MP).

In da Costa et al. [30], a different axiomatization is given, but it is essentially the same as ours. There, the postulates for implication are different. Namely, although $(\rightarrow_1)$ and $(\rightarrow_3)$ are the same (although the naming differs), the remaining axiom is:

$(A \rightarrow B) \rightarrow ((A \rightarrow (B \rightarrow C)) \rightarrow (A \rightarrow C))$

It is well known that there are many ways to axiomatize the implicational fragment of classical logic C. In the absence of negation, we need the so-called *Pierce's law* ($\rightarrow_3$) for C.

In ($\neg_1$), ($\neg_2$), ($\neg_3$), F and G are complex formulas. In general, without this restriction on F and G, these are not sound rules due to the fact that they are not admitted in annotated logics.

da Costa et al. [30] fuses (τ_1) and (τ_2) as the single axiom in conjunctive form. But, we separate it in two axioms for our purposes. Also there is a difference in the final axiom. They present it for infinite lattices as

$$A \rightarrow p_{\lambda_j} \text{ for every } j \in J, \text{ then } A \rightarrow p_\lambda, \text{ where } \lambda = \bigvee_{j \in J} \lambda_j.$$

If τ is a finite lattice, this is equivalent to the form of (τ_2).

As usual, we can define a *syntactic consequence relation* in $P\tau$. Let Γ be a set of formulas and G be a formula. Then, G is a syntactic consequence of Γ, written $\Gamma \vdash G$, iff there is a finite sequence of formulas $F_1, F_2, \ldots, F_n$, where F_i belongs to Γ, or F_i is an axiom ($1 \leq i \leq n$), or F_j is an immediate consequence of the previous two formulas by ($\rightarrow_4$). This definition can extend for the transfinite case in which n is an ordinal number. If $\Gamma = \emptyset$, i.e. $\vdash G$, G is a *theorem* of $P\tau$.

Let Γ, Δ be sets of formulas and A, B be formulas. Then, the consequence relation $\vdash$ satisfies the following conditions.

1. if $\Gamma \vdash A$ and $\Gamma \subset \Delta$ then $\Delta \vdash A$.
2. if $\Gamma \vdash A$ and $\Delta, A \vdash B$ then $\Gamma, \Delta \vdash B$.
3. if $\Gamma \vdash A$, then there is a finite subset $\Delta \subset \Gamma$ such that $\Delta \vdash A$.

In the Hilbert system above, the so-called *deduction theorem* holds.

Theorem 5.8 (Deduction theorem) *Let Γ be a set of formulas and A, B be formulas. Then, we have:*

$$\Gamma, A \vdash B \ \Rightarrow \ \Gamma \vdash A \rightarrow B.$$

The following theorem shows some theorems related to strong negation.

Theorem 5.9 *Let A and B be any formula. Then,*

1. $\vdash A \vee \neg_* A$
2. $\vdash A \rightarrow (\neg_* A \rightarrow B)$
3. $\vdash (A \rightarrow B) \rightarrow ((A \rightarrow \neg_* B) \rightarrow \neg_* A)$

From Theorems 5.9, 5.10 follows.

Theorem 5.10 *For arbitrary formulas A and B, the following hold:*

1. $\vdash \neg_*(A \wedge \neg_* A)$

2. $\vdash A \leftrightarrow \neg_* \neg_* A$

3. $\vdash (A \wedge B) \leftrightarrow \neg_*(\neg_* A \vee \neg_* B)$

4. $\vdash (A \rightarrow B) \leftrightarrow (\neg_* A \vee B)$

5. $\vdash (A \vee B) \leftrightarrow \neg_*(\neg_* A \wedge \neg_* B)$

Theorem 5.10 implies that by using strong negation and a logical connective other logical connectives can be defined as in classical logic. If $\tau = \{t, f\}$, with its operations appropriately defined, we can obtain classical propositional logic in which $\neg_*$ is classical negation.

Now, we provide some formal results of $P\tau$ including completeness and decidability.

Lemma 5.9 *Let p be a propositional variable and $\mu, \lambda, \theta \in |\tau|$. Then, the following hold:*

1. $\vdash p_{\lambda \vee \mu} \rightarrow p_\lambda$

2. $\vdash p_{\lambda \vee \mu} \rightarrow p_\mu$

3. $\lambda \geq \mu \ and \ \lambda \geq \theta \ \Rightarrow \vdash p_\lambda \rightarrow p_{\mu \vee \theta}$

4. $\vdash p_\mu \rightarrow p_{\mu \wedge \theta}.$

5. $\vdash p_\theta \rightarrow p_{\mu \wedge \theta}.$

6. $\lambda \leq \mu \ and \ \lambda \leq \theta \ \Rightarrow \vdash p_{\mu \wedge \theta}$

7. $\vdash p_\mu \leftrightarrow p_{\mu \vee \mu}, \ \vdash p_\mu \leftrightarrow p_{\mu \wedge \mu}$

8. $\vdash p_{\mu \vee \lambda} \leftrightarrow p_{\lambda \vee \mu}, \ \vdash p_{\mu \wedge \lambda} \leftrightarrow p_{\lambda \wedge \mu}$

9. $\vdash p_{(\mu \vee \lambda) \vee \theta} \vee \rightarrow p_{\mu \vee (\lambda \vee \theta)}, \ \vdash p_{(\mu \wedge \lambda) \wedge \theta} \vee \rightarrow p_{\mu \wedge (\lambda \wedge \theta)}$

10. $p_{(\mu \vee \lambda) \wedge \mu} \rightarrow p_\mu, \ p_{(\mu \wedge \lambda) \vee \mu} \rightarrow p_\mu$

11. $\lambda \leq \mu \ \Rightarrow \vdash p_{\lambda \vee \mu} \rightarrow p_\mu$

12. $\lambda \vee \mu = \mu \ \Rightarrow \vdash p_\mu \rightarrow p_\lambda$

13. $\mu \geq \lambda \ \Rightarrow \forall \theta \in |\tau| \ (\vdash p_{\mu \vee \theta} \rightarrow p_{\lambda \vee \theta} \ and \ \vdash p_{\mu \wedge \theta} \rightarrow p_{\lambda \wedge \theta})$

14. $\mu \geq \lambda \ and \ \theta \geq \varphi \ \Rightarrow \vdash p_{\mu \vee \theta} \rightarrow p_{\lambda \vee \varphi} \ and \ p_{\mu \wedge \theta} \rightarrow p_{\lambda \wedge \varphi}$

15. $\vdash p_{\mu \wedge (\lambda \vee \theta)} \rightarrow p_{(\mu \wedge \lambda) \vee (\mu \wedge \theta)}, \ \vdash p_{\mu \vee (\lambda \wedge \theta)} \rightarrow p_{(\mu \vee \lambda) \wedge (\mu \vee \theta)}$

16. $\vdash p_\mu \wedge p_\lambda \leftrightarrow p_{\mu \wedge \lambda}$

17. $\vdash p_{\mu \vee \lambda} \rightarrow p_\mu \vee p_\lambda$

Example 5.1 Consider the complete lattice $\tau = N \cup \{\omega\}$, where N is the set of natural numbers. The ordering on τ is the usual ordering on ordinals, restricted to the set τ. Consider the set $\Gamma = \{p_0, p_1, p_2, \ldots\}$, where $p_\omega \notin \Gamma$. It is clear that $\Gamma \vdash p_\omega$, but an infinitary deduction is required to establish this.

Definition 5.12 $\overline{\Delta} = \{A \in \mathbf{F} \mid \Delta \vdash A\}$

Definition 5.13 Δ is said to be *trivial* iff $\overline{\Delta} = \mathbf{F}$ (i.e., every formula in our language is a syntactic consequence of Δ); otherwise, Δ is said to be *non-trivial*. Δ is said to be *inconsistent* iff there is some formula A such that $\Delta \vdash A$ and $\Delta \vdash \neg A$; otherwise, Δ is *consistent*.

From the definition of triviality, the next theorem follows:

Theorem 5.11 Δ *is trivial iff* $\Delta \vdash A \wedge \neg A$ *(or* $\Delta \vdash A$ *and* $\Delta \vdash \neg_* A$*) for some formula* A.

Theorem 5.12 *Let* Γ *be a set of formulas,* A, B *be any formulas, and* F *be any complex formula. Then, the following hold.*

1. $\Gamma \vdash A$ *and* $\Gamma \vdash A \rightarrow B \Rightarrow \Gamma \vdash B$
2. $A \wedge B \vdash A$
3. $A \wedge B \vdash B$
4. $A, B \vdash A \wedge B$
5. $A \vdash A \vee B$
6. $B \vdash A \vee B$
7. $\Gamma, A \vdash C$ *and* $\Gamma, B \vdash C \Rightarrow \Gamma, A \vee B \vdash C$
8. $\vdash F \leftrightarrow \neg_* F$
9. $\Gamma, A \vdash B$ *and* $\Gamma, A \vdash \neg_* B \Rightarrow \Gamma \vdash \neg_* A$
10. $\Gamma, A \vdash B$ *and* $\Gamma, \neg_* A \vdash B \Rightarrow \Gamma \vdash B$.

Note here that the counterpart of Theorem 5.12 (10) obtained by replacing the occurrence of $\neg_*$ by $\neg$ is not valid.

Now, we are in a position to prove the soundness and completeness of $P\tau$. Our proof method for completeness is based on maximal non-trivial set of formulas; see Abe [1] and Abe and Akama [6]. da Costa et al. [30] presented another proof using Zorn's Lemma.

Theorem 5.13 (Soundness) *Let* Γ *be a set of formulas and* A *be any formula.* $\mathcal{A}\tau$ *is a sound axiomatization of* $P\tau$*, i.e., if* $\Gamma \vdash A$ *then* $\Gamma \models A$.

For proving the completeness theorem, we need some theorems.

Theorem 5.14 *Let* Γ *be a non-trivial set of formulas. Suppose that* τ *is finite. Then,* Γ *can be extended to a maximal (with respect to inclusion of sets) non-trivial set with respect to* $\mathbf{F}$.

Theorem 5.15 *Let Γ be a maximal non-trivial set of formulas. Then, we have the following:*

1. *if A is an axiom of $P\tau$, then $A \in \Gamma$*
2. *$A, B \in \Gamma$ iff $A \wedge B \in \Gamma$*
3. *$A \vee B \in \Gamma$ iff $A \in \Gamma$ or $B \in \Gamma$*
4. *if $p_\lambda, p_\mu \in \Gamma$, then $p_\theta \in \Gamma$, where $\theta = max(\lambda, \mu)$*
5. *$\neg^k p_\mu \in \Gamma$ iff $\neg^{k-1} p_{\sim\mu} \in \Gamma$, where $k \geq 1$*
6. *if $A, A \to B \in \Gamma$, then $B \in \Gamma$*
7. *$A \to B \in \Gamma$ iff $A \notin \Gamma$ or $B \in \Gamma$*

Theorem 5.16 *Let Γ be a maximal non-trivial set of formulas. Then, the characteristic function χ of Γ, that is, $\chi_\Gamma \to \mathbf{2}$ is the valuation function of some interpretation $I : \mathbf{P} \to |\tau|$.*

Here is the completeness theorem for $P\tau$.

Theorem 5.17 (Completeness) *Let Γ be a set of formulas and A be any formula. If τ is finite, then $\mathcal{A}\tau$ is a complete axiomatization for $P\tau$, i.e., if $\Gamma \models A$ then $\Gamma \vdash A$.*

The decidability theorem also holds for finite lattice.

Theorem 5.18 (Decidability) *If τ is finite, then $P\tau$ is decidable.*

The completeness does not in general hold for infinite lattice. But, it holds for special case.

Definition 5.14 (*Finite annotation property*) Suppose that Γ be a set of formulas such that the set of annotated constants occurring in Γ is included in a finite substructure of τ (Γ itself may be infinite). In this case, Γ is said to have the *finite annotation property*.

Note that if τ' is a substructure of τ then τ' is closed under the operations $\sim, \vee$ and $\wedge$. One can easily prove the following from Theorem 5.17.

Theorem 5.19 (Finitary Completeness) *Suppose that Γ has the finite annotation property. If A is any formula such that $\Gamma \vdash A$, then there is a finite proof of A from Γ.*

Theorem 5.19 tells us that even if the set of the underlying truth-values of $P\tau$ is infinite (countably or uncountably), as long as theories have the finite annotation property. The completeness result applied to them, i.e., $\mathcal{A}\tau$ is complete with respect to such theories.

In general, when we consider theories that do not possess the finite annotation property, it may be necessary to guarantee completeness by adding a new infinitary inference rule (ω-rule), similar in spirit to the rule used by da Costa [24] in order to cope with certain models in a particular family of infinitary language. Observe that for such cases a desired axiomatization of $P\tau$ is not finitary.

From the classical result of compactness, we can state a version of the compactness theorem.

Theorem 5.20 (Weak Compactness) *Suppose that Γ has the finite annotation property. If A is any formula such that $\Gamma \vdash A$, then there is a finite subset Γ' of Γ such that $\Gamma' \vdash A$.*

Annotated logics $P\tau$ provide a general framework, and can be used to reasoning about many different logics. Below we present some examples.

The set of truth-values $FOUR = \{t, f, \bot, \top\}$, with $\neg$ defined as: $\neg t = f, \neg f = t, \neg\bot = \bot, \neg\top = \top$. Four-valued logic based on $FOUR$ was originally due to Belnap [20, 21] to model internal states in a computer.

Subrahmanian [45] formalized an annotated logic with $FOUR$ as a foundation for paraconsistent logic programming; also see Blair and Subrahmanian [22].

Their annotated logic may be used for reasoning about inconsistent knowledge bases. For example, we may allow logic programs to be finite collections of formulas of the form:

$$(A : \mu_0) \leftrightarrow (B_1 : \mu_1)\& \cdots \&(B_n : \mu_n)$$

where A and B_i $(1 \le i \le n)$ are atoms and μ_j $(0 \le j \le n)$ are truth-values in $FOUR$.

Intuitively, such programs may contain "intuitive" inconsistencies–for example, the pair

$$((p : f), (p : t))$$

is inconsistent. If we append this program to a consistent program P, then the resulting union of these two programs may be inconsistent, even though the predicate symbols p occurs nowhere in program P.

Such inconsistencies can easily occur in knowledge based systems, and should not be allowed to trivialize the meaning of a program. However, knowledge based systems based on classical logic cannot handle the situation since the program is trivial.

In Blair and Subrahmanian [22], it is shown how the four-valued annotated logic may be used to describe this situation. Later, Blair and Subrahmanian's annotated logic was extended as *generalized annotated logics* by Kifer and Subrahmanian [37].

There are also other examples which can be dealt with by annotated logics. The set of truth-values $FOUR$ with negation defined as boolean complementation forms an annotated logic.

The unit interval $[0, 1]$ of truth-values with $\neg x = 1 - x$ is considered as the base of annotated logic for qualitative or fuzzy reasoning. In this sense, probabilistic and fuzzy logics could be generalized as annotated logics.

The interval $[0, 1] \times [0, 1]$ of truth-values can be also used for annotated logics for evidential reasoning. Here, the assignment of the truth-value (μ_1, μ_2) to proposition p may be thought of as saying that the degree of belief in p is μ_1, while the degree of disbelief is μ_2. Negation can be defined as $\neg(\mu_1, \mu_2) = (\mu_2, \mu_1)$.

Note that the assignment of $[\mu_1, \mu_2]$ to a proposition p by an interpretation I does not necessarily satisfy the condition $\mu_1 + \mu_2 \leq 1$. This contrasts with probabilistic reasoning. Knowledge about a particular domain may be gathered from different experts (in that domain), and these experts may different views.

Some of these views may lead to a "strong" belief in a proposition; likewise, other experts may have a "strong" disbelief in the same proposition. In such a situation, it seems appropriate to report the existence of conflicting opinions, rather than use ad-hoc means to resolve this conflict.

5.3   Predicate Annotated Logics $Q\tau$

As mentioned above, da Costa et al. [30] investigated propositional annotated logics $P\tau$, and suggested their predicate extension $Q\tau$ (also denoted QT). We can look at the detailed formulation of $Q\tau$ in da Costa et al. [26]; also see Abe [1].

Predicate annotated logics $Q\tau$ can be formalized as a two-sorted first-order logic. We repeat some definitions below. $\tau = \langle|\tau|, \leq, \sim\rangle$ is some arbitrary, but fixed complete lattice, with the ordering $\leq$ and the operator $\sim:|\tau|\to|\tau|$. The bottom element of this lattice is denoted by $\bot$, and top element is denoted by $\top$.

The language L^τ of $Q\tau$ is a first-order language without equality. Abe [1] introduced equality into $Q\tau$.

Definition 5.15 (*Symbols*) Primitive symbols are the following:

1. Logical connectives: $\wedge$ (conjunction), $\vee$ (disjunction), $\to$ (implication), and $\neg$ (negation)

2. Individual variables: a denumerably infinite set of variable symbols

3. Individual constants: an arbitrary family of constant symbols

4. Quantifiers: $\forall$ (for all) and $\exists$ (exists)

5. Function symbols: for each natural number $n > 0$, a collection of function symbols of arity n

6. Annotated predicate symbols: for any natural number $n \geq 0$, and any $\lambda \in \tau$, a family of annotated predicate symbols p_τ

7. Parentheses: (and)

Here, $\forall$ is called the *universal quantifier* and $\exists$ the *existential quantifier*. We define the notion of *term* as usual. Given an annotated predicate symbol p_λ of arity n and n terms $t_1, \ldots, t_n$, an *annotated atom* is an expression of the form $p_\lambda(t_1, \ldots, t_n)$.

Definition 5.16 (*Formulas*) Formulas are defined as follows:

1. An annotated atom is a formula.
2. If F is a formula, then $\neg F$ is a formula.
3. If F and G are formulas, then $F \wedge G$, $F \vee G$, $F \rightarrow G$ are formulas.
4. If F is a formula and x is an individual variable, then $\forall x F$ and $\exists x F$ are formulas.

Definition 5.17 (*Hyper-literal and complex formulas*) Hyper-literal and complex formulas are defined as follows. A formula of the form $\neg^k p_\mu(t_1, \ldots, t_n)$ $(k \geq 0)$ is called a *hyper-literal*. A formula which is not a hyper-literal is called a *complex formula*

As in $P\tau$, we may also use the formulas of the form $A \leftrightarrow B$ and $\neg_* A$ in $Q\tau$. Here, $\leftrightarrow$ denotes the *equivalence* and $\neg_*$ *strong negation*, respectively. We can also introduce the *equality*, denoted $=$, into $Q\tau$. If t and s are terms, then $s = t$ is also a formula. $s = t$ is read "s and t are equal".

Now, we describe a semantics for $Q\tau$, which is a variant of the semantics for standard first-order logic.

Definition 5.18 (*Interpretation*) An *interpretation* I for the language L^τ of $Q\tau$ consists of a non-empty set, denoted by $dom(I)$, and called the *domain*, together with

1. a function η_I that maps constants of L^τ to $dom(I)$
2. a function ζ_I that assigns, to each function symbol f of arity n in L^τ, a function from $(dom(I))^n$ to $dom(I)$
3. a function χ_I that assigns, to each predicate symbol of arity n in L^τ, a function from $(dom(I))^n$ to τ.

Definition 5.19 (*Variable assignment*) Suppose I is an interpretation for L^τ. Then, a *variable assignment* v for L^τ with respect to I is a map from the set of variables symbols of L^τ to $dom(I)$.

Definition 5.20 (*Denotation*) The *denotation* $d_{I,v}(t)$ of a term t with reference to an interpretation I and variable assignment v is defined inductively as follows:

1. If t is a constant symbol, then $d_{I,v}(t) = \eta(t)$.
2. If t is a variable symbol, then $d_{I,v}(t) = v(t)$.
3. If t is a function symbol, then $d_{I,v}(t) = \zeta(f)(d_{I,v}(t_1), \ldots, d_{I,v}(t_n))$.

Definition 5.21 (*Truth relation*) Let I and v be an interpretation of L^τ and a variable assignment with reference to I, respectively. We also suppose that A is an ordinary atom, and that F, G and H are any formulas whatsoever. Then, the *truth relation* $I, v \models A$, saying that A is true with reference to an interpretation I and variable assignment v, is defined as follows:

1. $I, v \models p_\mu(t_1, \ldots, t_n)$ iff $\chi_I(p)(d_{I,v}(t_1), \ldots, d_{I,v}(t_n)) \geq \mu$
2. $I, v \models \neg^k A_\mu$ iff $I, v \models \neg^{k-1} A_{\sim\mu}$
3. $I, v \models F \wedge G$ iff $I, v \models F$ and $I, v \models G$
4. $I, v \models F \vee G$ iff $I, v \models F$ or $I, v \models G$
5. $I, v \models F \rightarrow G$ iff $I, v \not\models F$ or $I, v \models G$
6. $I, v \models \neg F$ iff $I, v \not\models F$, where F is not a hyper-literal
7. $I, v \models \exists x H$ iff for some variable assignment v' such that for all variables y different from x, $v(y) = v'(y)$, we have that $I, v' \models H$
8. $I, v \models \forall x H$ iff for all variable assignments v' such that for all variables y different from x, $v(y) = v'(y)$, we have that $I, v' \models H$
9. $I \models H$ iff for all variable assignments v associated with I, $I, v \models H$

The equality $s = t$ is interpreted as follows:

$$I, v \models s = t \text{ iff } d_{I,v}(s) = d_{I,v}(t)$$

Here, $=$ at the right hand side of 'iff' denotes the equality symbol in the meta-language, and it reads classically. We could also introduce annotated quality $=_\lambda$ as a binary annotated atom. However, we do not go into details here.

We can define the notions of validity, model and semantic consequence as in Sect. 5.2. Let $\Gamma \cup \{H\}$ be a set of formulas. We write $\models H$, and say that H is *valid* (in $Q\tau$) if, for every interpretation I, $I \models H$. If $I \models A$ for each $A \in \Gamma$, I is a *model* of Γ. We say that H is a *semantic consequence* of Γ iff for any interpretation I such that $I \models G$ for all $G \in \Gamma$, it is the case that $I \models F$.

The following lemmas concerns the properties of $\models$, whose proofs are immediate from the corresponding proof in the previous chapter.

Lemma 5.10 *For any complex formula A and B and any formula F, the valuation v satisfies the following:*

1. $\models A \leftrightarrow B$ iff $\models A \rightarrow B$ and $\models B \rightarrow A$
2. $\not\models (A \rightarrow A) \wedge \neg(A \rightarrow A)$
3. $\models \neg_* A$ iff $\not\models A$
4. $\models \neg F \leftrightarrow \neg_* F$

Lemma 5.11 *Let $p_\mu(t_1, \ldots, t_n)$ be an annotated atom and $\mu, \lambda \in |\tau|$. Then, we have:*

1. $\models p_\perp(t_1, \ldots, t_n)$
2. $\models p_\mu(t1, \ldots, t_n) \rightarrow p_\lambda(t_1, \ldots, t_n)$, $\mu \geq \lambda$
3. $\models \neg^k p_\mu(t_1, \ldots, t_n) \leftrightarrow \neg^{k-1} p_{\sim\mu}(t_1, \ldots, t_n)$, $k \geq 0$

Next, we show a Hilbert style axiomatization of $Q\tau$, called $\mathcal{A}$. In the formulation of the postulates of $\mathcal{A}$, the symbols A, B, C denote any formula whatsoever, F and G denote complex formulas, and P_λ is an annotated atom.

Postulates for $\mathcal{A}$ described in Abe [1] are as follows; also see da Costa et al. [26].

Postulates for $\mathcal{A}$

$(\rightarrow_1)$ $(A \rightarrow (B \rightarrow A)$
$(\rightarrow_2)$ $(A \rightarrow (B \rightarrow C)) \rightarrow ((A \rightarrow B) \rightarrow (A \rightarrow C))$
$(\rightarrow_3)$ $((A \rightarrow B) \rightarrow A) \rightarrow A$
$(\rightarrow_4)$ $A, A \rightarrow B / B$
$(\wedge_1)$ $(A \wedge B) \rightarrow A$
$(\wedge_2)$ $(A \wedge B) \rightarrow B$
$(\wedge_3)$ $A \rightarrow (B \rightarrow (A \wedge B))$
$(\vee_1)$ $A \rightarrow (A \vee B)$
$(\vee_2)$ $B \rightarrow (A \vee B)$
$(\vee_3)$ $(A \rightarrow C) \rightarrow ((B \rightarrow C) \rightarrow ((A \vee B) \rightarrow C))$
$(\neg_1)$ $(F \rightarrow G) \rightarrow ((F \rightarrow \neg G) \rightarrow \neg F)$
$(\neg_2)$ $F \rightarrow (\neg F \rightarrow A)$
$(\neg_3)$ $F \vee \neg F$
$(\exists_1)$ $A(t) \rightarrow \exists x A(x)$
$(\exists_2)$ $A(x) \rightarrow B / \exists x A(x) \rightarrow B$
$(\forall_1)$ $\forall x A(x) \rightarrow A(t)$
$(\forall_2)$ $A \rightarrow B(x) / A \rightarrow \forall x B(x)$
(τ_1) $p_\perp(a_1, \ldots, a_n)$
(τ_2) $\neg^k p_\lambda(a_1, \ldots, a_n) \leftrightarrow \neg^{k-1} p_{\sim\lambda}(a_1, \ldots, a_n)$
(τ_3) $p_\lambda(a_1, \ldots, a_n) \rightarrow p_\mu(a_1, \ldots, a_n)$, where $\lambda \geq \mu$
(τ_4) If $A \rightarrow p_{\lambda_j}(a_1, \ldots, a_n)$, then $A \rightarrow p_\lambda(a_1, \ldots, a_n)$ for every $j \in J$,
where $\lambda = \bigvee_{i=1}^{m} \lambda_i$

As τ is a complete lattice, the supremum in (τ_4) is well-defined. The postulates for quantifiers are subject to the usual restrictions. When τ is finite, (τ_4) can be replaced by the schema:

$$p_{\lambda_1}(a_1, \ldots, a_n) \wedge p_{\lambda_2}(a_1, \ldots, a_n) \wedge \cdots \wedge p_{\lambda_m}(a_1, \ldots, a_n) \rightarrow p_\lambda(a_1, \ldots, a_n),$$
where $\lambda = \bigvee_{i=1}^{m} \lambda_i$

Here, $(\rightarrow_4)$, $(\exists_4)$, $(\forall_4)$ and (τ_4) are regarded as rules of inference

Abe [1] also added the following three axioms for equality:

$(=_1)$ $x = x$
$(=_2)$ $x_1 = y_1 \rightarrow (\ldots \rightarrow (x_n = y_n \rightarrow f(x_1, \ldots, x_n) = f(y_1, \ldots, y_n)))$
$(=_3)$ $x_1 = y_1 \rightarrow (\ldots \rightarrow (x_n = y_n \rightarrow P(x_1, \ldots, x_n) \rightarrow P(y_1, \ldots, y_n)))$

Here, f and P are function symbol and predicate symbol, respectively.

As in $\mathcal{A}\tau$, we easily define the syntactic concepts related to $\mathcal{A}$; in particular the concepts of syntactic consequence $\vdash$ is defined in the normal way. We only note that the notion of deduction (proof) is not finitary if τ is infinite.

da Costa, Abe and Subrahmanian's axiomatization of $\mathcal{A}$ adopts different naming for postulates, but it is equivalent to the above axiomatization. The deduction theorem (Theorem 2.8) also holds for $\mathcal{A}$.

Theorem 5.21 *The following dualities of quantifiers hold:*

1. $\vdash \forall x A \leftrightarrow \neg_* \exists x \neg_* A$
2. $\vdash \exists x A \leftrightarrow \neg_* \forall x \neg_* A$

Here are some formal results of $Q\tau$. The first result is soundness of $Q\tau$.

Theorem 5.22 (Soundness) *Let $\Gamma \cup \{A\}$ be a set of formulas of $Q\tau$. Then, $\Gamma \vdash A$ (in $\mathcal{A}$) implies that $\Gamma \vdash A$, i.e., $\mathcal{A}$ is sound with respect to the semantics of $Q\tau$.*

The next result is completeness of $Q\tau$ in a restricted sense.

Theorem 5.23 (Completeness) *Let $\Gamma \cup \{A\}$ be a set of formulas of $Q\tau$. Then, if τ is finite or if $\Gamma \cup \{A\}$ possesses the finite annotation property, we have that $\Gamma \models A$ entails $\Gamma \vdash A$, i.e., $\mathcal{A}$ is complete with respect to the semantics of $Q\tau$.*

When τ is infinite, it seems that completeness can be obtained only by augmenting $\mathcal{A}$ with an extra infinitary rule.

$Q\tau$ belong to the class of non-classical logics, and they are paraconsistent and paracomplete. They have a weak negation $\neg$, but we can define the strong negation $\neg_*$, which is classical.

da Costa et al. [26] presented another axiomatization of $Q\tau$ with a different nature, which is obtained by adjoining to the classical first-order logic, a weak negation $\neg$ plus some extra convenient postulates.

Let $\mathcal{C}$ be an axiomatization of classical first-order logic (without equality), in which negation is denoted by $\sim$. The remaining primitives defined symbols of $\mathcal{C}$ are the same as the corresponding one of $Q\tau$. We also suppose that the atomic formulas of the language of $\mathcal{C}$ are annotated atoms, as in L^τ. Furthermore, we suppose that we have added to $\mathcal{C}$ a weak negation $\neg$.

We denote by $\mathcal{A}'$ the axiomatic system obtained from $\mathcal{C}$ by adding the axioms $(\neg_1), (\neg_2), (\neg_3), (\tau_1), (\tau_2), (\tau_3), (\tau_4)$, and the rule:

> If F and G are formulas such that G is obtained from F by the replacement of a sub-formula of the form $\neg_* A$ by $A \to (A \to A) \wedge \neg(A \to A)$ or by the replacement of a sub-formula of the latter from by one of the first form, then infer $F \leftrightarrow G$.

Theorem 5.24 $\mathcal{A}$ *and* $\mathcal{A}'$ *are equivalent; both characterize* $Q\tau$.

Theorem 5.24 reveals that annotated logic $Q\tau$ can be interpreted as an extension of classical first-order logic C. This fact seems interesting theoretically as well as practically.

Annotated logics can be used for various mathematical subjects. For example, it is possible to work out a set theory based on $Q\tau$. We will explore annotated set theory. For this purpose, we need the notion of normal structure, and need to define a fragment of $Q\tau$.

Definition 5.22 (*Normal structure*) Let X be a non-empty set. A *normal structure* based on X is a function $f : X \times X \to \tau$.

We denote by $Q\tau^2$ the logic $Q\tau$ obtained by suppressing all function symbols and all predicate symbols, with the exception of one predicate symbol of arity 2 (a binary predicate symbol) which we represent by $\in$. $A\tau^2$ is then a dyadic predicate calculus whose atoms are annotated by τ. An annotated atom of $Q\tau^2$ has the form $\in_\lambda (a, b)$, where a and b are terms and $\lambda \in \tau$. This atom will be written $a \in_\lambda b$.

Intuitively, $\in$ is the membership predicate symbol. The subscript λ denotes a "degree" of membership. A normal structure is basically just a first-order interpretation as defined earlier with the following differences. First, $Q\tau^2$ contains only one predicate symbol $\in$ associated with different members of τ. Second, the normal structures are the interpretations of $\in$.

Theorem 5.25 *$Q\tau^2$ is sound with respect to the semantics of normal structures. If τ is finite or we consider only sets of formulas sharing the finite annotation property, then $Q\tau^2$ is also complete.*

5.4   Curry Algebras

We can develop an algebraic semantics for $P\tau$. Algebraic semantics is mathematically more elegant than model-theoretic semantics. However, algebraic semantics for paraconsistent logics challenges standard formulation, since known techniques cannot be properly used. Abe [3] proposed Curry algebra $P\tau$ that algebraizes propositional annotated logics $P\tau$. Abe proved the completeness theorem for $P\tau$ with respect to the algebraic semantics.

In order to obtain algebraic versions of the majority of logical systems the procedure is the following: we define an appropriate equivalence relation in the set of formulas (e.g. identifying equivalent formulas in classical propositional logic), in such a way that the primitive connectives are compatible with the equivalence relation, i.e., a congruence.

The resulting quotient system is the algebraic structure linked with the corresponding logical system. By this process, Boolean algebra constitutes the algebraic version of classical propositional logic, Heyting algebra constitutes the algebraic version of intuitionistic propositional logic, and so on. Thus, the procedure is to formulate an algebraic semantics.

However, in some non-classical logics, it is not always clear what "appropriate" equivalence relation here can be; the non-existence of any significant equivalence relation among formulas of the calculus can also take place. This occurs, for instance, with some paraconsistent systems; see Mortensen [39]. Indeed, as pointed out by Eytan [32], even for classical logic, it may not always be convenient to apply these ideas.

Now, we give some basic definitions related to Curry algebras. In $P\tau$, we define $A \leq B$ by setting $\vdash A \to B$, and $A \equiv B$ by setting $A \leq B$ and $B \leq A$. Here, $\leq$ is a quasi-order and $\equiv$ is an equivalence relation, respectively. Let R be a set whose elements are denoted by x, y, z, x', y'.

Definition 5.23 (*Curry pre-ordered system*) A system $(R, \equiv, \leq)$ is called a *Curry pre-ordered system*, if

1. $\equiv$ is an equivalence relation on R
2. $x \leq x$
3. $x \leq y$ and $y \leq z$ imply $x \leq z$
4. $x \leq y, x' \equiv$ and $y' \equiv y$ imply $x' \leq y'$.

Definition 5.24 (*Pre-lattice*) A system $(R, \equiv, \leq)$ is called a *pre-lattice*, if $(R, \equiv, \leq)$ is a Curry pre-ordered system and

1. $\inf\{x, y\} \neq \emptyset$
2. $\sup\{x, y\} \neq \emptyset$.

We denote by $x \wedge y$ one element of the set of $\inf\{x, y\}$ and by $x \vee y$ one element of the set of $\sup\{x, y\}$.

Definition 5.25 (*Implicative pre-lattice*) A system $(R, \equiv, \leq)$ is called a *implicative pre-lattice*, if

1. $(R, \equiv, \leq)$ is a pre-lattice
2. $x \wedge (x \to y) \leq y$
3. $x \wedge y \leq z$ iff $x \leq y \to z$.

Definition 5.26 An *implicative pre-lattice* $(R, \equiv, \leq)$ is called *classic* if $(x \to y) \to x \leq y$ (Peirce's law).

As is obvious from the above definitions, a classic implicative pre-lattice is a pre-algebraic structure which can characterize positive classical propositional logic, i.e., classical propositional logic without negation. As is well known, Peirce's law corresponds to the law of excluded middle.

We are now ready to define a Curry algebra $P\tau$. Let S be a non-empty set and $\tau = (|\tau|, \leq)$ be a finite lattice with the operation $\sim:|\tau|\to|\tau|$. We denote by S^* the set of all pairs (p, λ), where $p \in S$ and $\lambda \in |\tau|$.

We now consider the set $S^* \cup \{\neg, \wedge, \vee, \to\}$. Let S^{**} be the smallest algebraic structure freely generated by the set $S^* \cup \{\neg, \wedge, \vee, \to\}$ by the usual algebraic method. Elements of S^{**} are classified in two categories: *hyper-literal elements* are of the form $\neg^k(p, \lambda)$ and *complex elements* are the remaining elements of S^{**}.

Now, we introduce the concept of a Curry algebra $P\tau$.

Definition 5.27 (*Curry algebra $P\tau$*) A *Curry algebra $P\tau$* (abbreviated by $P\tau$-algebra) is a structure $R\tau = (R, (\natural, \leq, \sim), \equiv, \to, \neg)$ and, for $p \in R, a \in R^*, x, y \in R^{**}$:

1. R^{**} is a classical implicative lattice with a greatest element 1
2. $\neg$ is a unary operator $\neg : R^{**} \to R^{**}$
3. $x \to y \leq (x \to \neg y) \to \neg x$
4. $x \leq \neg x \to a$
5. $p_\perp \equiv 1$
6. $x \vee \neg x \equiv 1$
7. $\neg^k(p, \lambda) \equiv \neg^{k-1}(p, \sim \lambda), k \geq 1$
8. If $\mu \leq \lambda$ then $(p, \mu) \leq (p, \lambda)$

9. $(p, \lambda_1) \wedge (p, \lambda_2) \wedge \cdots \wedge (p, \lambda_n) \leq (p, \lambda)$, where $\lambda = \bigvee_{i=1}^{n} \lambda_i$

One can easily see that a $P\tau$-algebra is distributive and has a greatest element as well as a first element.

Definition 5.28 Let x be an element of a $P\tau$-algebra. We put:

$$\neg_* x = x \to ((x \to x) \wedge \neg(x \to x))$$

In a $P\tau$-algebra, $\neg_* x$ is a Boolean complement of x, so both $x \vee \neg_* x \equiv 1$ and $x \wedge \neg_* x \equiv 0$ hold.

Theorem 5.26 *In a $P\tau$-algebra, the structure composed by the underlying set and by operations $\wedge, \vee$ and $\neg_*$ is a pre-Boolean algebra. If we pass to the quotient through the basic relation $\equiv$, we obtain a Boolean algebra in the usual sense.*

A *pre-Boolean algebra* is a partial preorder $(R, \leq)$ such that the quotient by the relation $\equiv$. Thus, by definition of $P\tau$-algebra, the mentioned structure is a pre-Boolean algebra.

In addition, replacing the class of equivalent formulas by a formula can produce a usual Boolean algebra in which the meet $\wedge$ is conjunction, the join $\vee$ is disjunction, and the complement is negation.

Definition 5.29 Let $(R, (\natural, \leq, \sim), \equiv, \leq, \to, \neg)$ be a $P\tau$-algebra, and $(R, (\natural, \leq, \sim), \equiv, \leq, \to, \neg_*)$ the Boolean algebra that is isomorphic to the quotient algebra of $(R, (\natural, \leq, \sim), \equiv, \leq, \to, \neg_*)$ by $\equiv$ is called the Boolean algebra *associated with* the $P\tau$-algebra.

Hence, we can establish the following first representation theorem for $P\tau$-algebra.

Theorem 5.27 *Any $P\tau$-algebra is associated with a field of sets. Moreover, any $P\tau$-algebra is associated with the field of sets simultaneously open and closed of a totally disconnected compact Hausdorff space.*

This is not the only way of extracting Boolean algebra out of $P\tau$-algebra. There is another natural Boolean algebra associated with a $P\tau$-algebra.

Definition 5.30 Let $(R, (\natural, \leq, \sim), \equiv, \leq, \rightarrow, \neg)$ be a $P\tau$-algebra. By RC we indicate the set of all complex elements of $(R, (\natural, \leq, \sim), \equiv, \leq, \rightarrow, \neg)$.

Then, the structure $(RC, (\natural, \leq, \sim), \equiv, \leq, \rightarrow, \neg)$ constitutes a pre-Boolean algebra which we call Boolean algebra *c-associated with* the $P\tau$-algebra $(R, (\natural, \leq, \sim), \equiv, \leq, \rightarrow, \neg)$. Thus, we obtain a second representation theorem for $P\tau$-algebra.

Theorem 5.28 *Any $P\tau$-algebra is c-associated with a field of sets. Moreover, any $P\tau$-algebra is c-associated with the field of sets simultaneously open and closed of a totally disconnected compact Hausdorff space.*

Theorems 5.27 and 5.28 show us that $P\tau$-algebra constitute interesting generalizations of the concept of Boolean algebra. There are some open questions related to these results. How many non-isomorphic Boolean algebra associated with a $P\tau$-algebra is there? How many non-isomorphic Boolean algebra c-associated with a $P\tau$-algebra is there? The answers to these questions can establish connections of associated and c-associated algebra.

Next, we show soundness and completeness of $P\tau$-algebras using the notion of filter and ideal of a $P\tau$-algebra.

Definition 5.31 (*Filter*) Let $(R, (\natural, \leq, \sim), \equiv, \leq, \rightarrow, \neg)$ be a $P\tau$-algebra. A subset F of R is called a *filter* if:

1. $x, y \in F$ imply $x \wedge y \in F$
2. $x \in F$ and $y \in R$ imply $x \vee y \in F$
3. $x \in F, y \in R$, and $x \equiv y$ imply $y \in F$.

Definition 5.32 (*Ideal*) Let $(R, (\natural, \leq, \sim), \equiv, \leq, \rightarrow, \neg)$ be a $P\tau$-algebra. A subset I of R is called an *ideal* if:

1. $x, y \in I$ imply $x \vee y \in I$
2. $x \in I$ and $y \in R$ imply $x \wedge y \in F$
3. $x \in I, y \in R$, and $x \equiv y$ imply $y \in F$.

Then, we have the following lemma whose proof is trivial.

Lemma 5.12 *Let $(R, (\natural, \leq, \sim), \equiv, \leq, \rightarrow, \neg)$ be a $P\tau$-algebra. A subset F of R is a filter iff:*

1. *$x, y \in F$ imply $x \wedge y \in F$*
2. *$x \in F, y \in R$, and $x \leq y$ imply $y \in F$*
3. *$x \in F, y \in R$, and $x \equiv y$ imply $y \in F$.*

A subset I of R is an ideal iff:

1. $x, y \in I$ *imply* $x \vee y \in I$
2. $x \in I, y \in R$, *and* $x \leq y$ *imply* $y \in I$
3. $x \in I, y \in R$, *and* $x \equiv y$ *imply* $y \in I$.

Filters are partially ordered by inclusion. Filters that are maximal with respect to this ordering are called *ultrafilters*. By the Ultrafilter Theorem, every filter in $P\tau$-algebra can be extended to an ultrafilter.

Theorem 5.29 *Let F be an ultrafilter in a $P\tau$-algebra. Then, we have:*

1. $x \wedge y \in F$ *iff* $x \in F$ *and* $y \in F$
2. $x \vee y \in F$ *iff* $x \in F$ *or* $y \in F$
3. $x \rightarrow y \in F$ *iff* $x \notin F$ *or* $y \in F$
4. *If* $p_{\lambda_1}, p_{\lambda_2} \in F$, *then* $p_\lambda \in F$, *where* $\lambda = \lambda_1 \vee \lambda_2$
5. $\neg^k p_\lambda \in F$ *iff* $\neg^{k-1} p_{\sim \lambda} \in F$
6. *If* $x, x \rightarrow y \in F$, *then* $y \in F$

Definition 5.33 *If* $R\tau_1 = (R_1, (|\tau_1|, \leq_1, \sim_1), \equiv_1, \leq_1, \rightarrow_1, \neg_1)$ *and* $R\tau_2 = (R_2, (|\tau_1|, \leq_2, \sim_2), \equiv_2, \leq_2, \rightarrow_2, \neg_2)$ *are two $P\tau$-algebras, then a homomorphism of $R\tau_1$ into $R\tau_2$ is a map f of R_1 into R_2 which preserves the algebraic operations, i.e., such that for $x, y \in R_1$:*

1. $x \leq_1 y$ *iff* $f(x) \leq_2 f(y)$
2. $f(x \rightarrow_1 y) \equiv_2 f(x) \rightarrow_2 f(y)$
3. $f(\neg_1 x) \equiv_2 \neg_2 f(x)$
4. *If* $x \equiv_1 y$, *then* $f(x) \equiv_2 f(y)$
5. f *is also extended to a homomorphism of* $(|\tau_1|, \leq_1, \sim_1)$ *into* $(|\tau_2|, \leq_2, \sim_2)$ *in an obvious way (i.e., for instance, $f(\sim_1 \lambda) = \sim_2 f(\lambda)$).*

Then, as in the classical case, we can present the following theorem:

Theorem 5.30 *Let $R\tau_1$ and $R\tau_2$ be two $P\tau$-algebras and f a homomorphism from $R\tau_1$ into $R\tau_2$. Then, the set $\{x \in R_1 \mid f(x) \equiv_2 1_2\}$ (the shell of f) is a filter and the set $\{x \in R_2 \mid f(x) \equiv_2 0_2\}$ (the kernel of f) is an ideal.*

Theorem 5.31 *If the shell of a homomorphism f of $P\tau$-algebra is an ultrafilter, then*

1. $f(x) \equiv 1$ *and* $f(y) \equiv 1$ *iff* $f(x \wedge y) = 1$
2. $f(x) \equiv 1$ *or* $f(y) \equiv 1$ *iff* $f(x \vee y) = 1$
3. $f(x) \equiv 0$ *or* $f(y) \equiv 1$ *iff* $f(x \rightarrow y) = 1$

Definition 5.34 Let **F** be the set of all formulas of the propositional annotated logic $P\tau$ and f a homomorphism from **F** (considered as a $P\tau$-algebra) into an arbitrary $P\tau$-algebra. We write $f \models \Gamma$, where Γ is a subset of **F**, if for each $A \in \Gamma$, $f(A) \equiv 1$. $\Gamma \models A$ means that for all homeomorphisms f from **F** into an arbitrary $P\tau$-algebra, if $f \models \Gamma$, then $f(A) \equiv 1$.

Based on the above results, we can establish algebraic soundness and completeness of the propositional annotated logic $P\tau$.

Theorem 5.32 (Soundness) *If A is a provable formula of $P\tau$, i.e., $\vdash A$, then $f(A) \equiv 1$ for any homomorphism f from **F** (considered as a $P\tau$-algebra) into an arbitrary $P\tau$-algebra.*

To prove completeness, we need the following theorem:

Theorem 5.33 *Let U be an ultrafilter in **F**. Then, there is a homomorphism f from **F** into $2 = \{0, 1\}$ such that the shell of f is U.*

Theorem 5.34 (Completeness) *Let **F** be the set of all formulas of the propositional annotated logic $P\tau$ and $A \in$ **F**. Suppose that $f(A) \equiv 1$ for any homomorphism f from **F** (considered as a $P\tau$-algebra) into an arbitrary $P\tau$-algebra. Then, A is a provable formula of $P\tau$, i.e., $\vdash A$.*

Theorem 5.34 gives an alternative completeness result of propositional annotated logics $P\tau$ using Curry algebras $P\tau$. Curry algebras can also be applied to the completeness proof of other paraconsistent logics.

5.5   Formal Issues

There are several important formal issues about annotated logics. *Annotated set theory* can be regarded as a generalization of classical set theory. The most convenient way to study normal structures is to start with a classical set theory, for instance, Zermelo-Fraenkel set theory ZF and to treat them inside ZF. If we proceed this way, then annotated set theory constitutes a natural and immediate extension of fuzzy set theory.

A model theory based annotated predicate logics can be formalized as classical model theory. It is shown that all classical results can be adapted to $Q\tau$. For example, Abe and Akama studied the ultraproduct method for $Q\tau$ in [5]. In fact, $Q\tau$ can provide a unified framework for paraconsistent model theory.

It is interesting to work out a proof theory for annotated logics. Indeed a Hilbert system for annotated logics has been developed, but we need other proof methods for practical applications. For example, a natural deduction formulation was explored in Akama et al. [14] and a tableau formulation was given in Akama et al. [13]. It is also possible to describe sequent calculi for annotated logics. A proof-theoretic study of annotated logics is of help to automated reasoning.

Annotated modal logics can be formalized by extending annotated logics with modal operators. Abe [2] proposed annotated modal logics $S5\tau$ whose modality can be interpreted S5. Akama and Abe [9] investigated annotated modal logics $K\tau$ which corresponds to the normal modal logic.

Annotated logics can be also extended to other modal logics, e.g. temporal, epistemic and deontic logic. Abe and Akama [7] annotated temporal logics for reasoning about inconsistencies in temporal systems.

We can employ annotated logics as a basis for uncertain reasoning. In other words, versions of annotated logics can be formalized as fuzzy, evidential or probabilistic logics. We mentioned these possibilities above.

Work on fuzzy reasoning in annotated logics may be found in Akama et al. [10, 12]. We also attempted to unify annotated and possibilistic logics in Akama and Abe [11].

5.6   Conclusions

We gave a general introduction to annotated logics, which are now considered as important paraconsistent systems. We surveyed propositional and predicate annotated logics with proof and model theory. As an algebraic semantics based on Curry algebras was reviewed. We also make some remarks on formal issues of annotated logics.

We now know many systems of paraconsistent logic, but no systems can provide a unified framework for real applications. Abe and his co-workers established real applications using annotated logics for many years. In this sense, annotated logics can be seen as one of the promising paraconsistent systems. Recent applications of annotated logics to several areas may be found in Abe [4].

Acknowledgments   We are grateful to the referee and J.M. Abe for useful comments.

References

1. Abe, J. M.: *On the Foundations of Annotated Logics* (in Portuguese). Ph.D. Thesis, University of São Paulo, Brazil (1992)
2. Abe, J.M.: On annotated modal logics. Mathematica Japonica **40**, 553–56 (1994)
3. Abe, J.M.: Curry algebra $P\tau$. Logique et Analyse **161-162-163**, 5–15 (1998)
4. Abe, J.M. (ed.): Paraconsistent Intelligent Based-Systems. Springer, Heidelberg (2015)
5. Abe, J.M., Akama, S.: Annotated logics $Q\tau$ and ultraproduct. Logique et Analyse **160**, 335–343 (1997) (published in 2000)
6. Abe, J.M., Akama, S.: On some aspects of decidability of annotated systems. In: Arabnia, H.R. (ed.) Proceedings of the International Conference on Artificial Intelligence, vol. II, pp. 789–795. CREA Press (2001)
7. Abe, J.M., Akama, S.: Annotated temporal logics $\Delta\tau$. In: Advances in Artificial Intelligence: Proceedings of IBERAIA-SBIA 2000, LNCS 1952, pp. 217–226. Springer, Berlin (2000)

8. Abe, J.M., Akama, S., Nakamatsu, K.: Introduction to Annotated Logics. Springer, Heidelberg (2015)
9. Akama, S., Abe, J.M.: Many-valued and annotated modal logics. In: Proceedings of the 28th International Symposium on Multiple-Valued Logic, pp. 114–119. Fukuoka (1998)
10. Akama, S., Abe, J.M.: Fuzzy annotated logics. In: Proceedings of IPMU'2000, pp. 504–508. Madrid, Spain (2000)
11. Akama, S., Abe, J.M.: The degree of inconsistency in paraconsistent logics. In: Abe, J.M., da Silva Filho, J.I. (eds.) Logic, Artificial Intelligence and Robotics, pp. 13–23. IOS Press, Amsterdam (2001)
12. Akama, S., Abe, J.M., Murai, T.: On the relation of fuzzy and annotated logics. In: Proceedings of ASC'2003, pp. 46–51. Banff, Canada (2003)
13. Akama, S., Abe, J.M., Murai, T.: A tableau formulation of annotated logics. In: Cialdea Mayer, M., Pirri, F. (eds.) Proceedings of TABLEAUX'2003, pp. 1–13. Rome, Italy (2003)
14. Akama, S., Nakamatsu, K., Abe, J.M.: A natural deduction system for annotated predicate logic. In: Knowledge-Based Intelligent Information and Engineering Systems: Proceedings of KES 2007—WIRN 2007, Part II, pp. 861–868. Lecture Notes on Artificial Intelligence, vol. 4693. Springer, Berlin (2007)
15. Anderson, A., Belnap, N.: Entailment: The Logic of Relevance and Necessity I. Princeton University Press, Princeton (1976)
16. Anderson, A., Belnap, N., Dunn, J.: Entailment: The Logic of Relevance and Necessity II. Princeton University Press, Princeton (1992)
17. Batens, D.: Dynamic dialectical logics. In: Priest, G., Routley, R., Norman, J. (eds.) Paraconsistent Logic: Essay on the Inconsistent, pp 187–217. Philosophia Verlag, München (1989)
18. Batens, D.: Inconsistency-adaptive logics and the foundation of non-monotonic logics. Logique et Analyse **145**, 57–94 (1994)
19. Batens, D.: A general characterization of adaptive logics. Logique et Analyse **173–175**, 45–68 (2001)
20. Belnap, N.D.: A useful four-valued logic. In: Dunn, J.M., Epstein, G. (eds.) Modern Uses of Multi-Valued Logic, pp. 8–37. Reidel, Dordrecht (1977)
21. Belnap, N.D.: How a computer should think. In: Ryle, G. (ed.) Contemporary Aspects of Philosophy, pp. 30–55. Oriel Press (1977)
22. Blair, H.A., Subrahmanian, V.S.: Paraconsistent logic programming. Theor. Comput. Sci. **68**, 135–154 (1989)
23. Carnielli, W.A., Coniglio, M.E., Marcos, J.: Logics of formal inconsistency. In: Gabbay, D., Guenthner, F. (eds.) Handbook of Philosophical Logic, 2nd edn, vol. 14, pp. 1–93 Springer, Heidelberg (2007)
24. da Costa, N.C.A.: α-models and the system T and T^*. Notre Dame J. Form. Logic **14**, 443–454 (1974)
25. da Costa, N.C.A.: On the theory of inconsistent formal systems. Notre Dame J. Form. Logic **15**, 497–510 (1974)
26. da Costa, N.C.A., Abe, J.M., Subrahmanian, V.S.: Remarks on annotated logic. Zeitschrift für mathematische Logik und Grundlagen der Mathematik **37**, 561–570 (1991)
27. da Costa, N.C.A., Alves, E.H.: A semantical analysis of the calculi C_n. Notre Dame J. Form. Logic **18**, 621–630 (1977)
28. da Costa, N.C.A., Henschen, L.J., Lu, J.J., Subrahmanian, V.S.: Automatic theorem proving in paraconsistent logics: foundations and implementation. In: Proceedings of the 10th International Conference on Automated Deduction, pp. 72–86, Springer, Berlin (1990)
29. da Costa, N.C.A., Subrahmanian: Paraconsistent logic as a formalism for reasoning about inconsistent knowledge. Artif. Intell. Med. **1**, 167–174 (1989)
30. da Costa, N.C.A., Subrahmanian, V.S., Vago, C.: The paraconsistent logic $P\mathcal{T}$. Zeitschrift für mathematische Logik und Grundlagen der Mathematik **37**, 139–148 (1991)
31. Dunn, J.M.: Relevance logic and entailment. In: Gabbay, D., Gunthner, F. (eds.) Handbook of Philosophical Logic, vol. III, pp. 117–224. Reidel, Dordrecht (1986)

32. Eytan, M.: Tableaux de Smullyan, ensebles de Hintikka et tour ya: un point de vue Algebriquem. Math. Sci. Hum. **48**, 21–27 (1975)
33. Jaśkowski, S.: Propositional calculus for contradictory deductive systems (in Polish). Studia Societatis Scientiarun Torunesis, Sectio A **1**, 55–77 (1948)
34. Jaśkowski, S.: On the discursive conjunction in the propositional calculus for inconsistent deductive systems (in Polish). Studia Societatis Scientiarun Torunesis, Sectio A **8**, 171–172 (1949)
35. Kifer, M., Lozinskii, E.L.: RI: a logic for reasoning with inconsistency. In: Proceedings of LICS4, pp. 253–262 (1989)
36. Kifer, M., Lozinskii, E.L.: A logic for reasoning with inconsistency. J. Autom. Reason. **9**, 179–215 (1992)
37. Kifer, M., Subrahmanian, V.S.: On the expressive power of annotated logic programs. In: Proceedings of the 1989 North American Conference on Logic Programming, pp. 1069–1089 (1989)
38. Kifer, M., Subrahmanian, V.S.: Theory of generalized annotated logic programming. J. Logic Program. **12**, 335–367 (1992)
39. Mortensen, C.: Every quotient algebra for C_1 is trivial. Notre Dame J. Formal Logic **21**, 694–700 (1980)
40. Priest, G., Routley, R., Norman, J. (eds.): Paraconsistent Logic: Essays on the Inconsistent. Philosopia Verlag, München (1989)
41. Priest, G.: Logic of paradox. J. Philos. Logic **8**, 219–241 (1979)
42. Priest, G.: Paraconsistent logic. In: Gabbay, D. Guenthner, F. (eds.) Handbook of Philosophical Logic, 2nd edn., pp. 287–393. Kluwer, Dordrecht (2002)
43. Priest, G.: *In Contradiction: A Study of the Transconsistent*, 2nd edn. Oxford University Press, Oxford (2006)
44. Routley, R., Plumwood, V., Meyer, R.K., Brady, R.: *Relevant Logics and Their Rivals*, vol. 1. Ridgeview, Atascadero (1982)
45. Subrahmanian, V.: On the semantics of quantitative logic programs. In: Proceedings of the 4th IEEE Symposium on Logic Programming, pp. 173–182 (1987)

Chapter 6
Paraconsistent Artificial Neural Network for Structuring Statistical Process Control in Electrical Engineering

João Inácio da Silva Filho, Clovis Misseno da Cruz, Alexandre Rocco, Dorotéa Vilanova Garcia, Luís Fernando P. Ferrara, Alexandre Shozo Onuki, Mauricio Conceição Mario and Jair Minoro Abe

Real applications of paraconsistent logic in technology sectors such as electrical engineering materialized from the initiative and selfless efforts of several scientists. Among these, we highlight Prof. Dr. Jair Minoro Abe, honorary member of the Research Group in Paraconsistent Logic Applications. This Chapter seeks to pay tribute and offer thanks to our colleague, the eminent researcher, Jair Minoro Abe, on his 60th birthday.

Abstract In this study, we present an algorithmic structure based on paraconsistent annotated logic (PAL) that can simulate the calculi of average values present in a dataset and detect the variations of the average using only PAL concepts. We call the

J.I. da Silva Filho (✉) · C.M. da Cruz · A. Rocco · D.V. Garcia · L.F.P. Ferrara ·
A.S. Onuki · M.C. Mario · J.M. Abe
Research Group in Paraconsistent Logic Applications, UNISANTA, Santa Cecília University, Oswaldo Cruz Street, 288, Santos City, SP CEP 11045-000, Brazil
e-mail: inacio@unisanta.br

C.M. da Cruz
e-mail: clovismisseno@gmail.com

A. Rocco
e-mail: a.rocco@terra.com.br

D.V. Garcia
e-mail: dora@unisanta.br

L.F.P. Ferrara
e-mail: lfpferrara@uol.com.br

A.S. Onuki
e-mail: shozost@yahoo.com.br

M.C. Mario
e-mail: cmario@unisanta.br

J.M. Abe
e-mail: jairabe@uol.com.br

J.M. Abe
Graduate Program in Production Engineering, ICET, Paulista University, São Paulo, Brazil

© Springer International Publishing Switzerland 2016   77
S. Akama (ed.), *Towards Paraconsistent Engineering*, Intelligent Systems Reference Library 110, DOI 10.1007/978-3-319-40418-9_6

structure as paraconsistent artificial neural network for extraction of moving average (PANnet$_{MovAVG}$). As an example of its application, we use PANnet$_{MovAVG}$ to assist in the analysis of a final product quality index related to electrical engineering. To obtain the final result, we applied PANnet$_{MovAVG}$ to simulate the statistical behavior of the Statistical Process Control (SPC) by comparing values obtained with a ranking that establishes quality index standards based on electrical power distribution. First, tests were conducted using data with random values to verify the behavior of PANnet$_{MovAVG}$ and to set the optimum number of algorithms to form an optimized computational structure. Then, we used a database with actual electric voltage values generated by an electrical power system of an electrical power utility grid in Brazil. In the various tests, PANnet$_{MovAVG}$ appropriately detected changes and identified variations of electric voltage in 220-V transmission lines. The results show that PANnet$_{MovAVG}$ can be used to construct an efficient architecture for determining and monitoring quality scores with applications in various areas of engineering, especially for detecting quality index in an electricity distribution network.

Keywords Paraconsistent annotated logic · Statistical process control · Electrical power system · Energy quality

6.1 Introduction

Today's markets are increasingly demanding, and a continuous monitoring of the production process to achieve high product quality is essential for the survival of businesses [13]. In addition to offering a high-quality product to satisfy consumers, some industrial sectors, especially those dealing with electrical power systems, must satisfy government regulatory agencies that periodically check product quality and can apply penalties to companies that do not satisfy quality targets imposed by law. To address these problems, research conducted in production engineering seeks relatively higher quality levels with the implementation of new techniques and innovative activities surrounding its products. The principal objective is to investigate new methods to monitor processes for maintaining the required product quality at relatively lower operating costs [12, 13, 17].

Statistics plays an important role in the area of quality control because its techniques and methodologies have been increasingly used and accepted in organizations for this task. Among statistical methods, statistical process control (SPC) is widely used for controlling and monitoring the quality of final products [12].

6.1.1 Statistical Process Control SPC

SPC can be defined as a set of procedural tools that is intended to indicate whether a process is optimally working when only common causes of variation are present.

SPC is used when a process is disordered and requires some type of corrective action, i.e., when there are special causes of variation [12, 13, 17].

The primary objective of SPC is the systematic reduction of variability in product characteristics that affect quality, and SPC causes this reduction through statistical analysis and identification of deviations that affect the quality of the final product delivered to the consumer. For correct operation, SPC has tools to maintain adequate performance and predictability, as well as to detect changes in the behavior of the process. Changes must be detected as soon as possible to enable SPC to ensure that appropriate corrective actions are taken and that the process can be corrected without excessive damage to production [15, 16].

6.1.1.1 Moving Average

In statistics, a moving average is a set of numbers, each of which is the average of the corresponding subset of a larger set of data [11, 13]. Therefore, a moving average (moving mean or running average) involves a calculation to analyze data by creating a sequence of averages of different subsets of a full dataset [11, 13, 15].

6.1.1.2 Extracting the Moving Average

The extraction process of a moving average can be performed as follows [11].

Given a sequence of numbers and a fixed subset size, the first element of the moving average is obtained by taking the average of the initial fixed subset of the number sequence. In the next step the subset is modified by "shifting forward", that is, by excluding the first number of the sequence and including the next number following the original subset in the sequence. This creates a new subset of numbers, which is then averaged. This process is repeated over the entire data sequence. The sequence of plot points connecting all of the averages (fixed) is what we call the moving average. Therefore, a moving average is a set of numbers, each of which is the average of the corresponding subset of a larger set of data [11, 13, 15].

6.1.1.3 Control Charts

A control chart is a set of points (samples), ordered in time, that are interpreted in terms of horizontal lines, the upper control limit (UCL), central line (CL), and lower control limit (LCL) [15, 16].

Control charts were developed by Shewhart [17] and are now the most important tools in the analysis of process variability in industrial environments [12]. Based on statistical techniques for the analysis of the dispersion of a random variable, it is possible to determine, with some degree of confidence, the upper and lower limits for which the random variable can have values. Within these limits the process is considered to be under statistical control [11, 12, 15, 17].

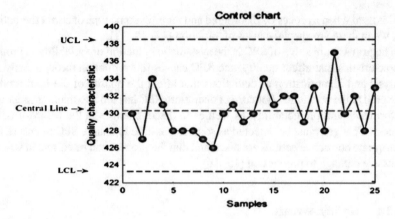

Fig. 6.1 Typical control chart

As can be seen in Fig. 6.1, the upper and lower lines represent the UCL and the LCL, respectively.

The CL relates to the mean of the values of the property studied. When all of the points shown are between the UCL and the LCL, the process is considered to be under statistical control [11, 12, 15, 17].

6.1.1.4 Application of Control Charts in SPC

SPC is a technique that makes it possible to monitor, analyze, predict, control, and improve the variability of a determined quality characteristic through the use of control charts. In practice, determining the limits of control is necessary for acquiring knowledge of the mean (μ) and the standard deviation (σ) of the process, when it is under statistical control. However, these values are not known with absolute precision; for this, we use an estimate created with samples of the process itself, to minimize the likelihood that there are extraneous causes [5, 11, 16].

The control chart limits $\bar{x}$ are calculated on the basis of the standard deviation of the sample average, as follows:

$$\sigma_{\bar{x}} = \frac{\sigma}{\sqrt{n}}$$

where

$\sigma_{\bar{x}}$ is the standard deviation of the mean of the process;
σ is the standard deviation of the process;
n is the number of samples.

The UCL is obtained by adding to the mean value μ the value of three standard deviations of the sample average:

$$UCL = \mu + 3\sigma_{\bar{x}}$$

The LCL is obtained by subtracting from the mean value μ the value of three standard deviations of the sample average:

$$LCL = \mu - 3\sigma_{\bar{x}}$$

The CL is the mean:

$$Center\,Line = \mu$$

The value of $\pm 3\sigma$ has a direct relation to the hypothesis testing of whether the sample average value is to be accepted as equal to the average of the process. In this respect, we have the following choices:

$H_0 : \mu = \bar{x}$, null hypothesis
$H_1 : \mu = \bar{x}$, alternate hypothesis

Using $\pm 3\sigma$ we have established confidence that 99.73 % of the average of the process is within the bounds of the control.

Although this technique is heavily applied, in some situations these graphs might not be sufficiently sensitive to detect small variations on the order of 1.5σ or less, because they do not take into consideration the values obtained historically, without considering the information provided by the sequence of points [16].

6.1.2 SPC Analysis

In performing its analysis, SPC uses the moving average and considers that the occurrence of special causes or those attributable to production changes the normal distribution in the average and/or standard deviation. These changes, which can exceed the lower and upper bounds and determine when the process was under statistical control, are detected and visualized in the control chart. Through the control charts, the system can find the trends through visualization of the resulting points that appear in the curve of the calculated average [11, 12, 15, 17].

Trend graphs are of paramount importance in the production process for decision-making regarding production quality. Thus, by checking the moving average, the trend of the curve, we can monitor the results over a period of time using control charts [5, 11, 15].

This procedure allows us to check continuously whether the process is performing as it is supposed to, or if it alternates, out of control, so that the necessary adjustments can be made in a timely manner.

SPC is a very powerful tool for decision-making, but there is an urgent need to make this method of monitoring and control more efficient both in analysis and in the extraction of knowledge. This increased efficiency can be achieved only through innovation in techniques applied in the analysis of tracked data and the systematization of these data to facilitate examination [11, 12, 14, 17].

6.1.2.1 SPC and Electrical Engineering

These features, using statistical calculations for the monitoring and control of variability, indicate that SPC can be used in applications in electrical engineering. In some studies, such as [5, 19], the authors used SPC techniques effectively for the analysis of monitoring and control of voltage levels in electrical power systems. However, owing to some factors, many of them inherent in statistical theories, other modes of SPC can be developed to improve their analysis and data interpretation methods. Following these considerations, we will present a simulation of statistical control through a configuration structured in paraconsistent artificial neural networks applied in electrical engineering [7, 8].

6.1.2.2 Electric Energy Quality Control

One of the most important applications of SPC in electrical engineering can be at electrical power distribution systems to evaluate the quality of electricity that is being made available to consumers. However, owing to the huge amount of data generated in an electrical power system, with incomplete and conflicting values forming uncertain databases, the statistical controls might not be efficient. To achieve the best level of quality with a method that is based on statistical analysis and with information from uncertain databases, work is needed with new algorithms based in non-classical logics. Only in this manner can the system of analysis offer responses to information represented by data that might be incomplete, ambiguous, or contradictory. Therefore, in this study we present an application of paraconsistent logic (PL), which, owing to its ability to provide effective treatment of contradictions, introduces new shapes to the analysis and interpretation of data used as indicators of levels of quality. Therefore, in this work, to work under the conditions of uncertainty that are generated by using incomplete data and that bring contradictory information to statistical calculations, we use algorithms based on a foundation of PL [1, 2, 7, 8].

To obtain values related to the average of the measurements of electrical magnitude data, we use a computational structure of algorithms called paraconsistent artificial neural cells (PANCells) [1, 2, 7, 8]. The configuration composed of PAN-Cell algorithms is called paraconsistent artificial neural network for extraction of moving average (PANnet$_{MovAVG}$) and is specially configured to calculate the moving average of the values of electrical magnitudes through iteration.

The principal concepts of PL, the algorithms of PANCell, and PANnet$_{MovAVG}$ itself will be examined in more detail in the following.

6.2 Paraconsistent Logic (PL)

PL is a non-classical logic whose primary feature is the rejection of the law of non-contradiction [7, 8]. The pioneers of PL are the Polish logician, J. Łukasiewicz, and the Russian philosopher, Vasilév [10], who simultaneously but independently in approximately 1910 suggested the possibility of the existence of a logic that did not use that principle of non-contradiction. The Brazilian logician N.C.A. da Costa was the first to publish work containing the initial systems of PL, including all of the logical levels, such as propositional and predicate calculi, as well as higher-order logics [7, 8, 10, 18].

6.2.1 Paraconsistent Annotated Logic (PAL)

PAL can be represented through a lattice of four vertices (Hasse lattice) in which a propositional sentence is accompanied by an evidence degree [1, 2, 7–9, 18].

The atomic formulae of PAL are of the form $p(\mu_1; \mu_2)$, where $(\mu_1; \mu_2) \in [0, 1]^2$, [0, 1] is the real unitary interval with the usual order relation, and p denotes a propositional variable. There is also an order relation defined on $[0, 1]^2 : (\mu_1; \mu_2) \leq (\lambda_1; \lambda_2) \leftrightarrow \mu_1 \leq \lambda_1$ and $\mu_2 \leq \lambda_2$. Such an ordered system constitutes a lattice that will be symbolized by τ.

6.2.1.1 PAL with Annotation of Two Values (PAL2v)

The atomic formulae of PAL, the first element (μ) of the pair (μ, λ), indicates the favorable evidence expressed by the proposition p, and the second element (λ) represents the unfavorable evidence expressed by p. Thus, the intuitive idea of the association of an annotation (μ, λ) to a proposition p is that the degree of favorable evidence in p is μ, and the degree of unfavorable (or contrary) evidence is λ. The pair (μ, λ) is called an annotation constant. In such a lattice, each of the annotation constants is related to a single extreme logical state of the proposition p, where ⊤= Inconsistent, t = True, F= False, and ⊥= Paracomplete or Indeterminate [1, 2, 7–9, 18]. The PAL2v lattice τ of four vertices (Hasse lattice) is shown in Fig. 6.2.

In the annotation constant we can consider that

(1, 0) → indicates total favorable evidence and no unfavorable evidence, an intuitive interpretation of true for proposition p;

(0, 1) → indicates zero favorable evidence and total unfavorable evidence, an intuitive interpretation of logical falsity for proposition p;

(1, 1) → indicates total favorable evidence and total unfavorable evidence, an intuitive interpretation of inconsistency for proposition p.

(0, 0) → indicates zero favorable evidence and no unfavorable evidence, an intuitive interpretation of paracompleteness or indetermination for proposition p.

Fig. 6.2 PAL2 lattice τ of
four vertices (Hasse Lattice)

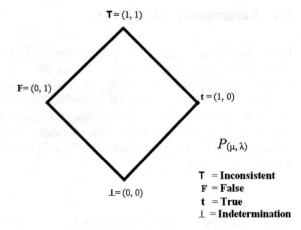

T = **Inconsistent**
F = **False**
t = **True**
⊥ = **Indetermination**

In this manner, a four-vertex lattice associated with a PAL with annotation of two
values (PAL2v) [7] can be represented as in Fig. 6.2.

Through linear transformations on the unit square in the Cartesian plane (USCP)
and the lattice associated with PAL2v, we can arrive at the equation of transformation:

$$T(X, Y) = (x - y, x + y - 1) \tag{6.1}$$

Referring to the relationship of the components of the transformation equation with
the usual nomenclature of PAL2v, we have:

$$x = \mu \rightarrow \text{favorable evidence degree}$$
$$y = \lambda \rightarrow \text{unfavorable evidence degree}$$

From the first term obtained in Eq. (6.1), this becomes $x - y = \mu - \lambda$, which we call
the degree of certainty (D_C). Therefore, the certainty degree [5] is calculated as

$$D_C = \mu - \lambda \tag{6.2}$$

The values, which belong to the set $\Re$ of real numbers, are in the closed interval
$[-1, +1]$, and they are on the horizontal axis of the lattice τ [5, 6, 16], which is called
the axis of degrees of certainty. When D_C results in $+1$, the logical state resulting
from a paraconsistent analysis is true (t), and when D_C results in -1, the logical state
resulting from the paraconsistent analysis is false (F).

From the second term Eq. (6.1), this becomes $x + y - 1 = \mu + \lambda - 1$, which is
called the degree of contradiction (D_{ct}) [7]. Therefore, the degree of contradiction
is obtained as

$$D_{ct} = \mu + \lambda - 1 \tag{6.3}$$

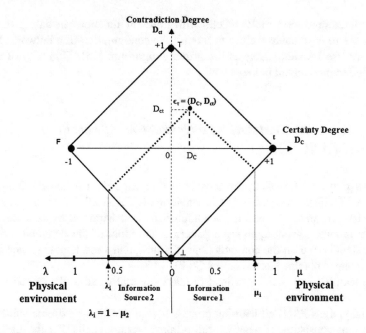

Fig. 6.3 Representation of the degrees of certainty and of contradiction in the PAL2v Lattice

Their values, which belong to the set $\Re$ of real numbers, vary in the closed interval $[-1, +1]$, and are on the vertical axis of the lattice τ [7], which is called the axis of degrees of contradiction. When D_{ct} results in $+1$, the logical state resulting from a paraconsistent analysis is Inconsistent (T), and when D_{ct} results in -1, the logical state resulting from the paraconsistent analysis is Indeterminate ($\perp$). Figure 6.3 shows the degrees of certainty and of contradiction in the PAL2v lattice.

6.3 Paraconsistent Artificial Neural Network (PANNet)

A PANNet consists of a computational structure formed of interconnections of algorithms based on PAL2v. The principle algorithms of a PANNet are called paraconsistent artificial neural cells (PANCells). The PANCells were projected to present characteristics able to model particular functions of an artificial neural network for decision-making [7].

[7, 8] presented a family of PANCells with components featuring several features. Each cell obtains a distinct functional response, which, when conveniently interconnected, comprises a PANNet with more specified functions in the control process.

The interconnection of PANCells with different functions in data processing allows one to have some facility to modify the configuration of a network. In this work, we use PANnet$_{\text{MovAVG}}$ as the primary algorithm [15]. This type of cell is described in more detail below.

6.3.1 Paraconsistent Artificial Neural Cell of Learning (LPANCell)

We now present LPANCell, the most important algorithm in the PANnet$_{\text{MovAVG}}$ structure [7]. Figure 6.4 shows a representation of LPANCell.

The learning of an LPANCell is accomplished through training by iteration, which consists in successively applying a pattern at the input of the favorable evidence degree signal (μ) until the contradictions diminish, and a resultant evidence degree equal to one is obtained as the output [7, 8].

The learning cells can be trained to learn any real value in the closed interval [0, 1].

Initially, the LPANCell learning process is shown with the extreme values zero or one, thus consisting of what we call primary sensorial cells. For the LPANCell learning process, a learning factor, F_L, is introduced, whose value can be adjusted externally. Depending on the value of F_L, faster or slower learning is provided to the LPANCell.

In the learning process, an equation for the values of the successive resultant evidence degree, $\mu_{E(k)}$, is considered until it acquires a value of one. Therefore, for an initial value of $\mu_{E(k)}$, the values $\mu_{E(k+1)}$ are obtained up to $\mu_{E(k+1)} = 1$.

Fig. 6.4 Representation of LPANCell

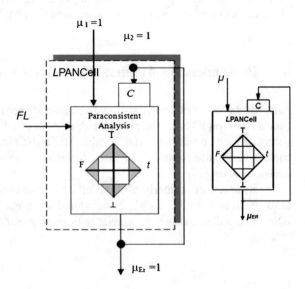

Considering the learning process of the truth pattern, the learning equation is obtained through the calculus of the resultant evidence degree equation:

$$\mu_{E(K+1)} = \frac{\{\mu_1 - (\mu_{E(K)C}) F_L\} + 1}{2} \tag{6.4}$$

where

$$\mu_{E(k)C} = 1 - \mu_{E(k)} \quad \text{and} \quad 0 \le F_L \le 1$$

The learning cell is considered completely trained when $\mu_{E(k+1)} = 1$.

For a learning process of the falsehood pattern, complementation on the favorable evidence degree is performed, and the equation becomes

$$\mu_{E(K+1)} = \frac{\{\mu_{1C} - (\mu_{E(K)C}) F_L\} + 1}{2} \tag{6.5}$$

where

$$\mu_{1C} = 1 - \mu_1 \quad \text{and} \quad 0 \le F_L \le 1$$

The learning cell is considered completely trained when $\mu_{E(k+1)} = 1$.

As seen from the calculus of the resultant degree of evidence equation $\mu_{E(k+1)}$, the higher the value, the faster the learning of the cell.

The learning factor F_L is a real value, in the closed interval [0, 1], and is attributed randomly by external adjustments. The flowchart and the learning algorithm are shown in Fig. 6.5 [7].

A simplified algorithm for learning any value between zero and one is shown below.

6.3.1.1   Algorithm of *L*PANCell

(For the Truth Pattern)

1- **Initial Condition**
   $\mu_1 = 1/2$ and $\mu_E = \mu_2 = 1/2$
2- **Enter the value of the learning factor**
   $F_L = C1$ */ Learning factor $0 \le F_L \le 1$ */
3- **Transform the evidence degree into an unfavorable evidence degree**
   $\lambda_2 = 1 - \mu_2$ */ unfavorable evidence degree $0 \le \lambda_2 \le 1$ */
4- **Enter the value of the evidence degree of input 1**
   $\mu_1 = 1$ */ evidence degree 1 */
5- **Compute the resultant evidence degree**
   $\mu_{E(K+1)} = \frac{\{\mu_1 - (\lambda_2)C_1\} + 1}{2}$

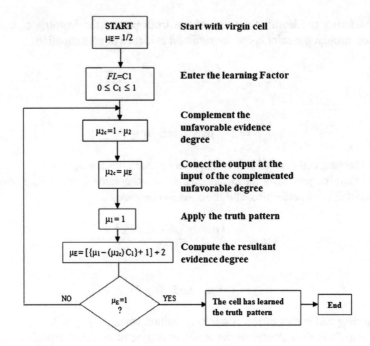

Fig. 6.5 Flowchart for the learning of the truth pattern by an *LPANCell*

6- **Consider the conditional**
    If $\mu_E \neq 1$ **Do** $\mu_E = \mu_2$ and **return** to step 3
7- **Stop**

6.4   Computational Structure PAL2v for Simulating SPC

In this work, we use PAL algorithms with PAL2v to form a computational structure able to determine SPC for obtaining a quality index of electrical energy [14]. The computational structure PAL2v has been developed with three main blocks.

6.4.1   *Extractor Block of Degrees of Evidence from z-Score*

This block is an algorithm that uses statistical concepts and PAL2v to extract degrees of evidence of data organized in a normal distribution represented by z-scores. The data are acquired from a historical database and also through real-time measurements.

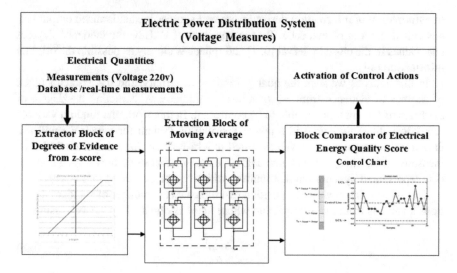

Fig. 6.6 Computational framework for analysis of electric energy quality factor (for steady state voltage item)

6.4.2 Extractor Block of Moving Average

This block is a configuration PANNet composed of six cells (PANCells) serially connected. This configuration of PANNet is used in the process of obtaining a value similar to the moving average of an electrical distribution system [14].

6.4.3 Block Comparator of Electrical Energy Quality Score

The moving average values obtained are continuously compared in a control chart that uses the upper and lower limits based on factors set out in the quality of electricity for the voltage. Figure 6.6 shows the computational structure used, with the highlights for the three blocks.

The next few items highlight each block comprising the computational structure PAL2v used in this work.

6.4.4 Operation of the Extractor Block of Evidence Degrees from z-Scores

The extraction of the evidence degrees is the first part of any computational arrangement for the application of PAL. This action consists in modeling that takes the transformation of information derived from real sources in standardized output values called degrees of evidence. By the concepts of PAL2v, the evidence degrees take values in the closed interval [0, 1] and reflect as closely as possible the original information [14].

In this analysis we seek the quality score of electricity, and the extractor block of degrees of evidence from z-scores has the function of extracting the degree of evidence of the random variable to be controlled. In this work, the random variable refers to voltage measured in a power distribution system for which we will use data from measurements on values stored in a history database and from real-time measurements. Thus, the evidence degrees extractor block operates from a mass of data represented by a normal distribution, which can be converted into z-scores.

The z-score is a measure of position that indicates the number of standard deviations a value is from the average, and whose value is obtained by the expression

$$z = \frac{x - \bar{x}}{\sigma} \tag{6.6}$$

where

x is the value of the variable
$\bar{x}$ is the arithmetic mean
σ is the standard deviation

Being dependent on the average and standard deviation of the variable x, the z values are within a range of $\pm 3z$ when the process is under statistical control.

As an example, if $x = -3$ in a system with and $\sigma = 1$, then $z = -3$ and $\mu = 0.25$. We note that the same holds if $x = 3$, $z = 3$, and $\mu = 0.75$.

Based on the previous example, however, changing the standard deviation to $\sigma = 0.5$, we have $z = -6$ and $\mu = 0$ for $x = -3$ and $z = 6$ and $\mu = 1$ for $x = 3$.

Thus, we can choose the upper and lower limits of the z-score, respectively, as 6 and -6.

In that manner, we can obtain the average shift represented by a normal curve of standard deviation $\pm 2\sigma$, as seen in Fig. 6.7.

As seen in Fig. 6.8, we can consider the normal curve to have zero average and double the standard deviation, i.e., $\sigma = 2$, without the values exceeding the maximum degree of evidence equal to unity.

To build the algorithm of the extractor of evidence degrees, we considered that the database of electrical tension constitutes a source of information that has the characteristics of a random x-variable with a normal distribution [14]. Thus, for the extraction of evidence (μ) of the random variable represented in measures of electrical voltage, a linear transformation function is used as in the expressions

$$\mu = \begin{cases} 0 \ if \ z < -0.6 \\ \frac{z}{12} + 0.5 \ if \ z \in [-0.6, +0.6] \\ 1 \ if \ z > +0.6 \end{cases} \tag{6.7}$$

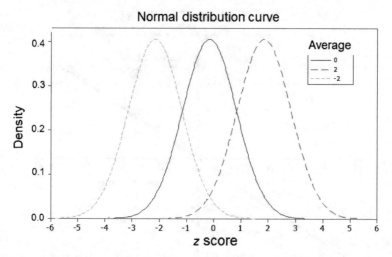

Fig. 6.7 Normal curve—average shifting

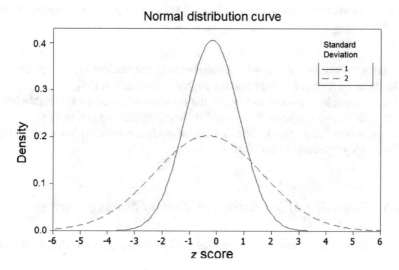

Fig. 6.8 Normal curve—increase in standard deviation

where

μ is the evidence degree.

z is the z-score.

We can see that in Eq. (6.6), if the random variable x is equal to the average, the z-score becomes equal to zero, and using Eq. (6.7) verifies that the degree of evidence μ assumes the value of 0.5. Similarly, when the variable x is equal to the average of the system, its value will always be represented by the degree of evidence equal to

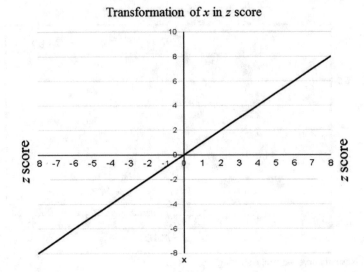

Fig. **6.9** Variable x transformed by the z-score to have the average zero and standard deviation equal to one

0.5, and there will be a direct relationship between the random variable represented in the measurements of voltage and the degree of evidence μ [14].

As an example, we show in Fig. 6.9 the transformation of an x-variable into z-scores with an average equal to zero and standard deviation equal to one.

The function that performs the extraction of evidence degree (μ) from the z-score is displayed graphically in Fig. 6.10.

6.4.5 Operation of the Extractor Block of Moving Average

The block that extracts the moving average is a $PANnet_{MovAVG}$ built with $LPANCells$ [7–9, 15].

After a few iterations, the degree of evidence resulting as output will be in the form of amplitudes that express the mean of the data applied in the input. Therefore, the extractor block of moving averages receives the signal of evidence (evidence degrees) from the extractor block of degrees of evidence of the z-score and provides in its output a value of the resulting evidence that represents a value similar to the moving average of the process.

The structure of $PANnet_{MovAVG}$ can consist of $nLPANCells$ [15], but depending on the number of components the process can be slow or fast. The assay procedure for the computational structure of optimized $PANnet_{MovAVG}$ is described below.

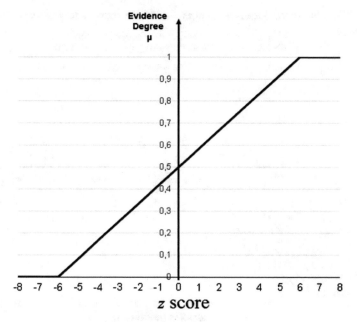

Fig. 6.10 Graph of degree of evidence extractor from z-score

6.4.5.1 PANnet$_{MovAVG}$ Optimization Procedures

The optimization process [15] initially involved a test in a single *LPANCell*, which was presented for its input a sequence of ones and zeroes, values that represent, respectively, the maximum degree of evidence and the minimal degree of evidence.

For these two symmetric values applied alternately, the value of the resulting evidence expected for a PANnet$_{MovAVG}$ with a single *LPANCell* is one half ($\mu_E = 0.5$), a value equivalent to the average expected statistically.

In this PANnet$_{MovAVG}$ configuration consisting of a single *LPANCell*, the output value was stabilized after about 20 samples (20 iterations), showing a cyclic variation between a minimum value of $\mu_E = 0.333333$ and a maximum value of $\mu_E = 0.666667$.

To verify the behavior of the resulting evidence degree in the output, other procedures were used with PANnet$_{MovAVG}$ composed of more cells connected serially. In the final process of optimization of PANnet$_{MovAVG}$, a configuration composed of 12 *LPANCells* was considered to simulate their serial arrangement.

Forty sequential samples (iterations) with alternating values of zero and one were introduced as the input of the first *LPANCell* for this final configuration [7, 14].

The results obtained with the PANnet$_{MovAVG}$ in the final arrangement of 12 *LPANCells* are shown in the graphs in Fig. 6.11 [14].

The results obtained in the simulation, as shown in Fig. 6.11, indicate that when the output range of the *LPANCell* is closer to the expected average, an increase

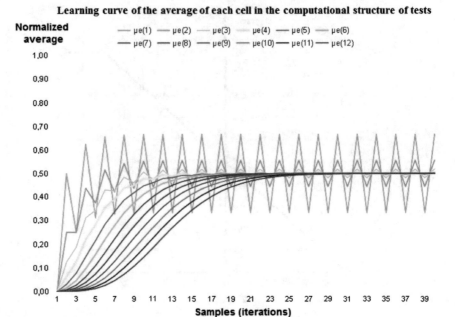

Fig. 6.11 Learning curves of the resulting evidence degrees in the output with the average of the *L*PANCells with serial interconnections (PANnet$_{MovAVG}$)

occurs in the number of *L*PANCells; however, for a greater number of *L*PANCells, the number of samples (iterations) must also be greater for the output value to remain around the mean.

6.4.5.2   The Final Configuration of PANnet$_{MovAVG}$

In the procedures of the tests that were completed with 12 *L*PANCells, it was found that some arrangements of *L*PANCells connected serially presented the best relationships between accuracy and response time, with the output showing a value close to the value of the expected average.

In this work the choice of optimized configuration was based in the error between the value of the evidence obtained and the expected average. Thus, we can establish a relationship between the error and the number of *L*PANCells used in the configuration of PANnet$_{MovAVG}$. Figure 6.12 shows a graph of the relationship between the percentage error and the number of cells used in PANnet$_{MovAVG}$ [14].

With these results the error can be determined by relating the number of cells used in the PANnet$_{MovAVG}$ configuration and the desired precision. The Eq. (6.8) [14] shows the relationship between the error and the number of cells used.

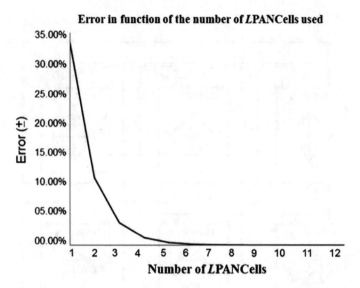

Error in function of the number of *LPANCells* used

Fig. 6.12 Percentage of residual error as a function of the number of *LPANCells* used

$$Error = \pm \frac{3^{-n_{lPANcell}}}{2} \tag{6.8}$$

where $n_{lPANcell}$ is the number of *LPANCells* used in the PANnet$_{MovAVG}$ configuration.

In the testing framework built with 12 *LPANCells*, we see that

(a) the arrangement of cells began to provide values around the mean from the twenty-first sample (iteration);
(b) for six *LPANCells*, the answer had its value stabilized only in the thirty-fourth sample (iteration);
(c) for six *LPANCells*, resulting values near $\mu = 0.49$ appeared from the sixteenth interaction;

By Eq. 6.8, for six cells, the error is $Error = \pm 0.000686$

Based on the study, and considering these results in conjunction with the response time of the setting of the serial cells, an arrangement with six *LPANCells* showed a performance with a low level of residual error, with an acceptable number of samples (iterations).

These considerations verify that a serial optimized arrangement of six *LPANCells* can be used for the determination of moving average and with conditions to acquire the arithmetic mean achieving convergence. A structure consisting of six *LPANCells* connected serially was built with PANnet$_{MovAVG}$ in this manner [14].

The complete PANnet$_{MovAVG}$ with six *LPANCells* used in its construction is shown in Fig. 6.13.

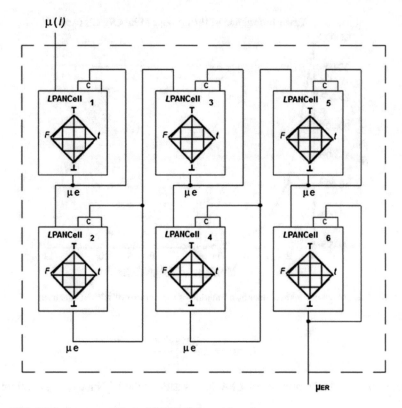

Fig. 6.13 PANnet$_{MovAVG}$ with six *L*PANCells in serial mode

After its structure was optimized, the PANnet$_{MovAVG}$ with six *L*PANCells was tested using simulators for normally distributed data, and the evidence degree of output obtained (μ_E) showed behavior compatible with the value of the moving average of the data applied in the input [7, 9, 14].

6.4.6 Operation of Block Comparator of Electric Energy Quality Score

In Brazil, the National Electric Energy Agency (ANEEL) [3] is a governmental regulatory agency that proposes resolutions or technical standards (NT-ANEEL) [4, 16] to which the electric energy distribution companies report good indices of quality to their consumers.

In general the NT-ANEEL rules [4] state that the quality of electric power in Brazil is based on

(a) quality of service and
(b) quality of the product.

The quality of the services provided is to be calculated by the determination of indicators of continuity and attendance times in emergency cases. In this study, we focus only on product quality, which is based on sampling criteria and on reference values that can be measured.

The reference values in product quality are considered to include the aspects dealing with a permanent or transitional regime, linked to the following factors:

(a) steady state voltage;
(b) power factor;
(c) harmonics;
(d) voltage imbalance;
(e) voltage fluctuation;
(f) voltage variations of short duration;
(g) frequency variation.

In this work, for verification of the quality of the electrical energy distributed by applying the concepts of PAL2v, we use only the criteria for "steady state voltage." In this manner, in the control chart we will use as UCL and LCL the sampling criteria of electrical energy quality defined by ANEEL's technical standard (NT-ANEEL) [4], as seen below.

6.4.6.1 Steady State Voltage

In this work, the analysis with the algorithms of PAL2v will indicate the quality of the product in relation to the reference values regarding the conformity of steady state voltage. For this we will use some terms of the technical standard (NT-ANEEL) that deal with quality levels of electrical energy regarding the behavior of permanent voltage, as in the conditions reported below.

- According to technical standard NT-ANEEL [4], the voltage in steady state operation must be accompanied throughout the distribution system, and the distribution company must establish resources and modern techniques for such monitoring.
  The company must act preventively, so that the values of electrical voltage in steady state remain within the appropriate standards.
- The steady state voltage must be evaluated through a set of readings obtained for proper measurement, in accordance with the methodology described for collective and individual indicators.

With respect to the reference values the technical standard NT-ANEEL [4] provides that

(a) the voltage values obtained by measurements must be compared to the reference voltage, which must be the nominal voltage, according to the voltage level of the connection point;
(b) the nominal values must be fixed on the basis of the levels of distribution system planning, so that there is compatibility with project levels of end-use electrical equipment;
(c) for each reference voltage, the associated readings are classified into three categories: appropriate, precarious, or critical, based on the remoteness of the read voltage value in relation to the reference voltage.

In the technical standard NT-ANEEL [4], the appropriate limits are established, critical to the voltage levels in voltage in steady state operation, the individual and collective compliance indicators of electrical voltage, measurement criteria, and deadlines for consumer compensation, if the voltage measurements exceed the limits of the indicators.

The service associated with the voltage readings must be classified according to the boundaries around the reference voltage (TR), such that

1. Reference voltage (TR);
2. Appropriate voltage range (TR $-$ ΔADINF, TR $+$ ΔADSUP);
3. Precarious voltage range (TR $+$ ΔADSUP, TR $+$ ΔADSUP $+$ ΔPRSUPorTR $-$ ΔADINF $-$ ΔPRINF, TR $-$ ΔADINF);
4. Critical voltage range ($>$TR $+$ ΔADSUP $+$ ΔPRSUP or $<$TR $-$ ΔADINF $-$ ΔPRINF).

Figure 6.14 shows the voltage levels with the band limits that determine quality in reference to values in steady state operation, according to ANEEL standards [4].

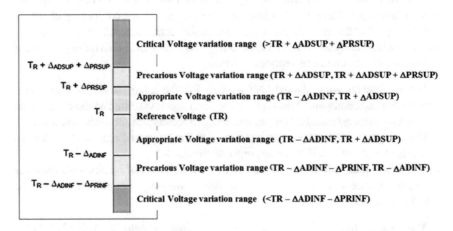

Fig. 6.14 Range of variation in a quality score that deals with voltage steady state operation

For the period of scouting the technical standard NT-ANEEL [4] specifies seven days with 10 min intervals between each reading with a total of 1,008 reads.

The individual indicators indices to be determined relate to the duration on the transgression of precarious voltage (DPV).

The DPV index is the ratio between the measurements in the precarious range and the total quantity of measurements; therefore, it is calculated as

$$DPV = \frac{nlp}{1008} \times 100\%$$

with a limit of 3 %
where nlp is the number of reads in the precarious range.

Duration of the transgression of Critical Voltage (DCV).

The DCV index is the ratio of the measurements in the critical range and the total number of measurements; therefore, it is calculated as

$$DCV = \frac{nlc}{1008} \times 100\%$$

with a limit of 0.5 %
where nlc is the number of reads on critical range.

The set of readings to generate individual indicators must be considered as the registry of 1,008 valid readings obtained in consecutive intervals of 10 min each. Each set of 1,008 valid readings composes a DPV indicator and a DCV indicator.

In this work, the assays were made using the reference voltage of 220 V. Table 6.1 shows the tension of meeting the specifications of the technical standard NT-ANEEL [4] for a study in low voltage of 220 V.

Based on Table 6.1, we show in Fig. 6.15 a score with the acceptable levels for the electrical voltage of 220 V used in some regions of Brazil. Therefore, in this study, this is the score used to compare the values of the moving average in the electric power system through the control chart.

6.4.6.2 Dataset

The data used in this work are offline and were obtained through the output bus voltage of a 220 V transformer in a substation installed in an electrical system in

Table 6.1 Connection points in nominal voltage equal to or less than 1 kV (220 V)

Electrical voltage of service to consumers (TR) (V)	Range of variation electric voltage reading (TL) (V)
Appropriate	$(202 \leq TL \leq 231)$
Precarious	$(191 \leq TL \leq 202)$
Critical	$(TL < 191 \text{ or } TL > 233)$

Fig. 6.15 Electrical voltage
levels with their range of
variation limits TR = 220 V
values in voltage steady state
operation

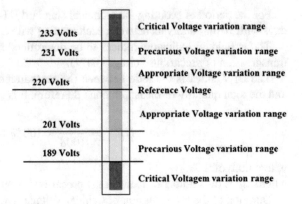

Brazil in which the measurements were taken during the period from 9/1/2007 to
4/30/2008. The database contains samples every 10 min and the use of such samples
was spaced to assure the non-existence of autocorrelation of data and thus facilitated
the exposure of results.

6.5   Results

The graph in Fig. 6.16 represents the results of the extraction of moving average
with the data extracted from the database used in this work. The upper and lower
limits of the average of the measured values of electrical voltage were established in
accordance with the rules of ANEEL [4].

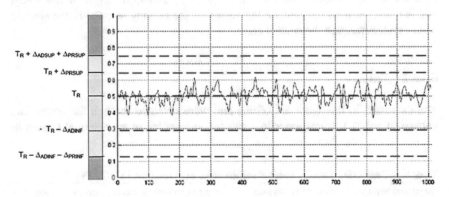

Fig. 6.16 Control chart showing the results of the extraction of the electrical energy quality index
from the comparison of NT-ANEEL standards [4] and values of the mobile average extracted with
the concepts of PAL2v

6.6 Conclusions

In this work, we presented an important application of PANnet in electrical engineering. We presented PANnet$_{MovAVG}$, an algorithmic structure based on PAL2v comprising an innovative method for obtaining a hybrid model combining a PAN-Net with the well-known statistical technique SPC. As we saw in this work, SPC provides the necessary tools for assessment and improvement of processes, products, and services in a robust and comprehensive manner, but such tools also depend on the quality of information and the manner in which the data are collected, processed, and interpreted. In practice, primarily in electrical engineering, data are collected using equipment subject to factors that generate uncertainty. Thus PANnet$_{MovAVG}$, containing PAL-based algorithms, can be a good alternative to provide highest efficiency in the control of the quality of electrical power distributed to consumers. Based on these results, new tests with different databases are currently being conducted to provide improvements in PANnet$_{MovAVG}$ performance for detecting moving averages and providing support to decision-making regarding the quality of electrical energy.

References

1. Abe, J.M., Akama, S., Nakamatsu, K.: Introduction to Annotated Logics—Foundations for Paracomplete and Paraconsistent Reasoning, Series Title Intelligent Systems Reference Library, Vol. 88. Publisher Springer International Publishing, Copyright Holder Springer International Publishing Switzerland, eBook ISBN 978-3-319-17912-4, doi:10.1007/978-3-319-17912-4, Hardcover ISBN 978-3-319-17911-7, Series ISSN 1868-4394, Edition Number 1, 190 pages (2015)
2. Abe, J.M.: Paraconsistent intelligent based-systems: new trends in the applications of paraconsistency, editor. Book Series: Intelligent Systems Reference Library, Vol. 94, 306 pages. Springer-Verlag, Germany (2015). ISBN 978-3-319-19721-0
3. BRAZIL-National Electric Energy Agency (Agência Nacional de Energia Elétrica)—ANEEL. In Portuguese: Atlas of Brazil's Electrical Energy. (Atlas de Energia Elétrica do Brasil). (2008). http://www.aneel.gov.br/arquivos/PDF/atlas_par1_cap2.pdf. Accessed 30 July 2015
4. BRAZIL-National Electric Energy Agency (Agência Nacional de Energia Elétrica)—ANEEL. In Portuguese: Procedures for the distribution of electricity in the National Electrical System (Procedimentos de Distribuição de Energia Elétrica no Sistema Elétrico Nacional—PRODIST) (2009). http://www.aneel.gov.br/arquivos/pdf/modulo8_24032006_srd.pdf. Accessed 20 May 2015
5. Corduas, M.: Bootstrapping moving average models. J. Italian Stat. Soc. 1, (2), 227–234 (1992). doi:10.1007/BF02589032 (Springer-Verlag)
6. Da Costa, N.C.A., Abe, J.M., Subrahmanian, V.S.: Remarks on annotated logic. Zeitschrift f. math. Logik und Grundlagen d. Math. 37, 561–570 (1991)
7. Da Silva Filho, J.I., Lambert-Torres, G., Abe, J.M.: Uncertainty Treatment Using Paraconsistent Logic: Introducing Paraconsistent Artificial Neural Networks, pp. 211, 328. Frontiers in Artificial Intelligence and Applications. IOS Press, Amsterdam, Netherlands (2010)
8. Da Silva Filho, J.I. et al.: Paraconsistent Logic Algorithms Applied to Seasonal Comparative Analysis with Biomass Data Extracted by the Fouling Process. Paraconsistent Intelligent-Based Systems: New Trends in the Applications of Paraconsistency: Intelligent Systems Reference Library, pp 131–152. Springer International Publishing AG, Switzerland (2015). doi:10.1007/978-3-319-19722-7

9. Da Silva Filho, J.I., Rocco, A., Mario, M.C., Ferrara, L.F.P.: Annotated Paraconsistent logic applied to an expert System Dedicated for supporting in an Electric Power Transmission Systems Re- Establishment. IEEE Power Engineering Society—PSC 2006, pp. 2212–2220. Atlanta, USA (2006). ISBN-1- 4244-0178-X
10. Dugan, R.C, McGranaghan, M.F., Santos, S., Beaty, H.W.: Electrical Power System Quality, 2nd edn., p. 528. McGraw Hill (2003)
11. Jacob, A.L., Pillai, S.K.: Statistical process control to improve coding and code review. IEEE Softw. **20**(3), 50–55 (2003)
12. Jelali, M.: Statistical Process Control. Control Performance Management in Industrial Automation Part of the series Advances in Industrial Control, pp. 209–217. Springer, London (2013). doi:10.1007/978-1-4471-4546-2_8
13. Montgomery, D.C.: Introduction to Statistical Quality Control, 4th edn. Wiley (2001). Young, M.: The Technical Writer's Handbook. University Science, Mill Valley, CA (1989)
14. Misseno da Cruz, C. et al.: Application of Paraconsistent Artificial Neural network in Statistical Process Control acting on voltage level monitoring in Electrical Power Systems. Intelligent System Application to Power Systems (ISAP), pp. 1–6. Porto–PT (2015). doi:10.1109/ISAP. 2015.7325579
15. Naikan, V.N.A.: Statistical Process. Control Handbook of Performability Engineering, pp. 187–201. Springer, London (2008). doi:10.1007/978-1-84800-131-2
16. Sancho, J., Pastor, J.J., Martínez, J., García M.A.: Evaluation of Harmonic Variability in Electrical Power Systems through Statistical Control of Quality and Functional Data Analysis. The Manufacturing Engineering Society International Conference, MESIC (2013)
17. Shewhart, W.A.: Economic Control of Quality of Manufactured Product, p. 501. Van Nostrand, New York (1931)
18. Subrahmanian, V.S.: On the semantics of quantitative logic programs. In: Proceedingsof 4th IEEE Symposium on Logic Programming. Computer Society Press, Washington DC (1987)
19. WESTERN ELECTRIC: Statistical Quality Control Handbook, 2nd edn. Western Electric Company, Indianapolis (1958)

Chapter 7
Programming with Annotated Logics

Kazumi Nakamatsu and Seiki Akama

Dedicated to Jair Minoro Abe for his 60th birthday

Abstract In this chapter, we survey paraconsistent annotated logic programs EVALPSN/bf-EVALPSN and their application to intelligent control, especially logical safety verification based control. We have already proposed a paraconsistent annotated logic program called EVALPSN. In EVALPSN, an annotation called an extended vector annotation is attached to each literal. For dealing with before-after relation between two time intervals, we also have introduced a new interpretation for extended vector annotations in EVALPSN, which is named before-after(bf)-EVALPSN. First, we review EVALPSN, and paraconsistent annotated logics PT and the basic annotated logic program are given as the formal background of EVALPSN/bf-EVALPSN with some simple examples. Then, EVALPSN is formally defined and its application to traffic signal control is described. We also introduce EVALPSN application to pipeline valve control with examples. Bf-EVALPSN is formally defined and its unique and useful reasoning rules are introduced with some examples. Last, we give some concluding remarks.

Keywords Annotated logic · Paraconsistent annotated logic programming · Intelligent control · EVALPSN · Bf-EVALPSN

K. Nakamatsu (✉)
University of Hyogo, 1-1-12 Shinzaike, Himeji 670-0092, Japan
e-mail: nakamatu@shse.u-hyogo.ac.jp

S. Akama
C-Republic, 1-20-1 Higashi-Yurigaoka, Asao-ku, Kawasaki 215-0012, Japan
e-mail: akama@jcom.home.ne.jp

© Springer International Publishing Switzerland 2016
S. Akama (ed.), *Towards Paraconsistent Engineering*, Intelligent Systems
Reference Library 110, DOI 10.1007/978-3-319-40418-9_7

7.1 Introduction

Paraconsistent logic is a logic capable of formalizing inconsistent but non-trivial theories. In standard logical systems like classical logic, inconsistency gives rise to triviality. In trivial theories, every formula is provable. However, for many applications we have to tolerate inconsistency. For this purpose, paraconsistent logic is suitable because it can deal with inconsistency in a framework of consistent logical systems.

It has been almost seven decades since the first paraconsistent logical system was proposed by Jaśkowsk [13]. It was three decades later that a family of paraconsistent logic called "annotated logic" was proposed by da Costa et al. [9, 50], which can deal with inconsistency by introducing many truth values called "annotations" into their syntax as attached information to atomic formulas. The paraconsistent annotated logic by da Costa et al. was originally developed from the viewpoint of logic programming by Subrahmanian et al. [8, 14, 49]. For details on annotated logic, the reader is advised to consult Abe et al. [1].

Furthermore, in order to deal with inconsistency and non-monotonic reasoning in a framework of annotated logic programming, ALPSN (Annotated Logic Program with Strong Negation) and its stable model semantics was developed by Nakamatsu and Suzuki [17]. It has been shown that ALPSN can deal with some non-monotonic reasonings such as default logic [47], autoepistemic logic [16] and a non-monotonic Assumption Based Truth Maintenance System (ATMS) [10] in a framework of annotated logic programming [18, 36, 37].

Even though ALPSN can deal with non-monotonic reasoning such as default reasoning and conflicts can be represented as paraconsistent knowledge in it, it is difficult and complicated to deal with reasoning to resolve conflicts in ALPSN. On the other hands, it is known that defeasible logic can deal with conflict resolving in a logical way [6, 42, 43], although defeasible logic cannot deal with inconsistency in its syntax and its inference rules are too complicated to be implemented easily.

In order to deal with conflict resolution and inconsistency in a framework of annotated logic programming, a new version of ALPSN, VALPSN (Vector Annotated Logic Program with Strong Negation) that can deal with defeasible reasoning and inconsistency was also developed by Nakamatsu [22]. Moreover, it has been shown that VALPSN can be applied to conflict resolution in various systems [19–21]. It also has been shown that VALPSN provides a computational model of defeasible logic [6, 7].

Later, VALPSN was extended to EVALPSN (Extended VALPSN) by Nakamatsu et al. [23, 24] to deal with deontic notions (obligation, permission, forbiddance, etc.) and defeasible deontic reasoning [44, 45]. Recently, EVALPSN has been applied to various kinds of safety verification and intelligent control, for example, railway interlocking safety verification [27], robot action control [25, 28, 29, 38], safety verification for air traffic control [26], traffic signal control [30], discrete event control [31–33] and pipeline valve control [34, 35].

Considering the intelligent safety verification for process control, there is an occasion in which the safety verification for process order control is significant. For example, suppose a pipeline network in which two kinds of liquids, nitric acid and caustic soda are used for cleaning the pipelines. If those liquids are processed continuously and mixed in the same pipeline by accident, explosion by neutralization would be caused. In order to avoid such a dangerous accident, the safety for process order control should be strictly verified in a formal way such as EVALPSN.

However, it seems to be a little difficult to utilize EVALPSN for verifying process order control as well as the safety verification for each process in process control. We have already proposed a new version of EVALPSN called bf(before-after)-EVALPSN that can deal with before-after relations between two time intervals (processes) by using two sorts of reasoning rules. One is named Basic Before-after reasoning rule and another is Transitive Before-after reasoning rule [39–41].

This chapter mainly reviews EVALPSN, bf-EVALPSN and their applications to intelligent control based on their logical reasoning. As far as we know there seems to be no other efficient computational tool that can deal with the real-time intelligent safety verification for process order control than bf-EVALPSN.

This chapter is organized as follows: firstly, in Sect. 7.1, the background and development of the paraconsistent annotated logic programs are overviewed.

In Sect. 7.2, paraconsistent annotated logics and their logic programs are introduced as the background knowledge of EVALPSN/bf-EVALPSN, moreover EVALPSN are formally recapitulated.

In Sect. 7.3, the traffic signal control system based on EVALPSN deontic defeasible reasoning and its simple simulation results by the cellular automaton method are given as an application of EVALPSN to intelligent control.

In Sect. 7.4, a simple brewery pipeline network is introduced and its pipeline process control system based on EVALPSN valve safety verification is shown to be as an application of EVALPSN to intelligent control systems.

In Sect. 7.5, the basic concepts of bf-EVALPSN are introduced and bf-EVALPSN itself is formally defined, furthermore, an application of bf-EVALPSN to real-time intelligent safety verification for process order control is presented with the pipeline examples in Sect. 7.4.

In Sect. 7.6, we review reasoning of before-after relations in bf-EVALPSN and a unique and useful inference method of before-after relations in bf-EVALPSN, which can be implemented as a bf-EVALPSN called "transitive bf-inference rules", is introduced with a simple example.

Lastly, in Sect. 7.7, we give some concluding remarks.

7.2 Paraconsistent Annotated Logic Program

This section is concerned with the formal background of the paraconsistent annotated logic program EVALPSN. For more details of EVALPSN; see [39]. We assume that the reader is familiar with the basic knowledge of classical logic and logic programming [15].

7.2.1 Paraconsistent Annotated Logic *PT*

In order to understand EVLPSN and its reasoning we introduce Paraconsistent Annotated Logics *PT* [9]. Here, we briefly give the syntax and semantics for propositional paraconsistent annotated logics *PT* proposed by da Costa et al. [9].

Generally, a truth value called an *annotation* is attached to each atomic formula explicitly in paraconsistent annotated logic, and the set of annotations constitutes a complete lattice. We introduce a paraconsistent annotated logic *PT* with the four valued complete lattice $\mathcal{T}$.

Definition 2.1 The primitive symbols of *PT* are:

1. propositional symbols: $p, q, \ldots, p_i, q_i, \ldots$;
2. each member of $\mathcal{T}$ is an *annotation constant* (we may call it simply an annotation);
3. the connectives and parentheses: $\wedge, \vee, \rightarrow, \neg, (,)$.

Formulas are defined recursively as follows:

1. if p is a propositional symbol and $\mu \in \mathcal{T}$ is an annotation constant, then $p:\mu$ is an *annotated atomic formula (atom)*;
2. if F, F_1, F_2 are formulas, then $\neg F, F_1 \wedge F_2, F_1 \vee F_2, F_1 \rightarrow F_2$ are formulas.

We suppose that the four-valued lattice in Fig. 7.1 is the complete lattice $\mathcal{T}$, where annotations t and f may be intuitively regarded as truth values *true* and *false*, respectively. It may be comprehensible that annotations $\perp$, t, f and $\top$ correspond to the truth values $*, T, F$ and *TF* in Visser [51] and **None, T, F**, and **Both** in Belnap [5], respectively. Moreover, the complete lattice $\mathcal{T}$ can be viewed as a bi-lattice in which the vertical direction $\overrightarrow{\perp \top}$ indicates *knowledge amount* ordering and the horizontal direction $\overrightarrow{ft}$ does *truth* ordering [11]. We use the symbol $\leq$ to denote the ordering in terms of knowledge amount (the vertical direction $\overrightarrow{\perp \top}$) over the complete lattice $\mathcal{T}$, and the symbols $\perp$ and $\top$ are used to denote the bottom and top elements, respectively. In the paraconsistent annotated logic *PT*, each annotated atomic formula can be interpreted epistemically, for example, $p:t$ may be interpreted epistemically as "the proposition p is known to be true".

There are two kinds of negation in the paraconsistent annotated logic *PT*, one of them is called *epistemic negation* and represented by the symbol $\neg$ (see Definition 2.1). The epistemic negation in *PT* followed by an annotated atomic formula is defined as a mapping between elements of the complete lattice $\mathcal{T}$ as follows:

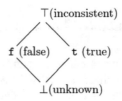

Fig. 7.1 The 4-valued complete lattice $\mathcal{T}$

$$\neg(\bot) = \bot, \quad \neg(t) = f, \quad \neg(f) = t, \quad \neg(\top) = \top.$$

As shown in the above mapping the epistemic negation maps annotations to themselves without changing the knowledge amounts of annotations. Furthermore, the epistemic negation followed by an annotated atomic formula can be eliminated by the mapping. For instance, the knowledge amount of annotation t is the same as that of annotation f as shown in the complete lattice $\mathcal{T}$, and we have the epistemic negation,[1]

$$\neg(p:t) = p:\neg(t) = p:f,$$

which shows that the knowledge amount in terms of the proposition p cannot be changed by the epistemic negation mapping. There is another negation called *ontological(strong) negation* that is defined by using the epistemic negation.

Definition 2.2 (*Strong Negation*)

Let F be any formula,

$$\sim F =_{def} F \to ((F \to F) \wedge \neg(F \to F)).$$

The epistemic negation in Definition 2.2 is not interpreted as a mapping between annotations since it is not followed by an annotated atomic formula. Therefore, the strongly negated formula $\sim F$ can be interpreted so that if the formula F exists, the contradiction $((F \to F) \wedge \neg(F \to F))$ is implied. Usually, the strong negation is used for denying the existence of a formula following it.

The semantics for the paraconsistent annotated logics PT is defined.

Definition 2.3 Let ν be the set of all propositional symbols and $\mathcal{F}$ be the set of all formulas. An interpretation I is a function,

$$I : \nu \longrightarrow \mathcal{T}.$$

To each interpretation I, we can associate the valuation function such that

$$v_I : \mathcal{F} \longrightarrow \{0, 1\},$$

which is defined as:

1. let p be a propositional symbol and μ an annotation,

$$v_I(p:\mu) = 1 \;\; \textbf{iff} \;\; \mu \leq I(p),$$
$$v_I(p:\mu) = 0 \;\; \textbf{iff} \;\; \mu \nleq I(p);$$

---

[1] An expression $\neg p:\mu$ is conveniently used for expressing a negative annotated literal instead of $\neg(p:\mu)$ or $p:\neg(\mu)$.

2. let A and B be any formulas, and A not an annotated atom,

$$v_I(\neg A) = 1 \ \textbf{iff} \ v_I(A) = 0,$$
$$v_I(\sim B) = 1 \ \textbf{iff} \ v_I(B) = 0;$$

other formulas $A \rightarrow B, A \wedge B, A \vee B$ are valuated as usual.

We provide an intuitive interpretation for strongly negated annotated atoms with the complete lattice $\mathcal{T}$. For example, the strongly negated literal $\sim(p:\mathtt{t})$ implies the knowledge "p is false ($\mathtt{f}$) or unknown ($\bot$)" since it denies the existence of the knowledge that "p is true ($\mathtt{t}$)". This intuitive interpretation is provided by Definition 2.3 as follows: if $v_I(\sim(p:\mathtt{t})) = 1$, we have $v_I(p:\mathtt{t}) = 0$ and for any annotation $\mu \in \{\bot, \mathtt{f}, \mathtt{t}, \top\} \leq \mathtt{t}$, we have $v_I(p:\mu) = 1$, therefore, we obtain that $\mu = \mathtt{f}$ or $\mu = \bot$.

7.2.2 EVALPSN (Extended Vector Annotated Logic Program with Strong Negation)

Generally, an annotation is explicitly attached to each literal in paraconsistent annotated logic programs as well as the paraconsistent annotated logic PT. For example, let p be a literal, μ an annotation, then $p:\mu$ is called an *annotated literal*. The set of annotations constitutes a complete lattice.

An annotation in EVALPSN has a form of $[(i, j), \mu]$ called an *extended vector annotation*. The first component (i, j) is called a *vector annotation* and the set of vector annotations, which constitutes a complete lattice,

$$\mathcal{T}_v(n) = \{(x, y) | 0 \leq x \leq n, 0 \leq y \leq n, x, y \text{ and } n \text{ are integers}\}$$

shown by the Hasse's diagram as $n = 2$ in Fig. 7.2.

The ordering($\preceq_v$) of the complete lattice $\mathcal{T}_v(n)$ is defined as follows: let (x_1, y_1), $(x_2, y_2) \in \mathcal{T}_v(n)$,

Fig. 7.2 Lattice $\mathcal{T}_v(2)$ and lattice $\mathcal{T}_d$

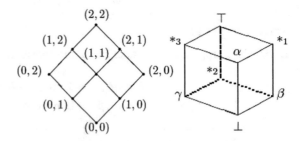

$$(x_1, y_1) \preceq_v (x_2, y_2) \text{ iff } x_1 \leq x_2 \text{ and } y_1 \leq y_2.$$

For each extended vector annotated literal $p : [(i, j), \mu]$, the integer i denotes the amount of positive information to support the literal p and the integer j denotes that of negative one. The second component μ is an index of fact and deontic notions such as obligation, and the set of the second components constitutes the following complete lattice,

$$\mathcal{T}_d = \{\bot, \alpha, \beta, \gamma, *_1, *_2, *_3, \top\}.$$

The ordering($\preceq_d$) of the complete lattice $\mathcal{T}_d$ is described by the Hasse's diagram in Fig. 7.2. The intuitive meaning of each member in $\mathcal{T}_d$ is

> $\bot$(unknown),
>
> α(fact), β(obligation), γ(non-obligation),
>
> $*_1$(fact and obligation),
>
> $*_2$(obligation and non-obligation),
>
> $*_3$(fact and non-obligation),   and
>
> $\top$(inconsistency).

The complete lattice $\mathcal{T}_d$ is a quatro-lattice in which the direction $\overrightarrow{\bot\top}$ measures *knowledge amount*, the direction $\overrightarrow{\gamma\beta}$ does *deontic truth*, the direction $\overrightarrow{\bot*_2}$ does *deontic knowledge amount* and the direction $\overrightarrow{\bot\alpha}$ does *factuality*.

For example, annotation β(obligation) can be intuitively interpreted to be more obligatory than annotation γ(non-obligation), and annotations $\bot$(no knowledge) and $*_2$(obligation and non-obligation) are deontically neutral, that is to say, it cannot be said whether they represent obligation or non-obligation.

The complete lattice $\mathcal{T}_e(n)$ of extended vector annotations is defined as the product,

$$\mathcal{T}_v(n) \times \mathcal{T}_d.$$

The ordering($\preceq_e$) of the complete lattice $\mathcal{T}_e(n)$ is also defined as follows: let $[(i_1, j_1), \mu_1], [(i_2, j_2), \mu_2] \in \mathcal{T}_e$,

$$[(i_1, j_1), \mu_1] \preceq_e [(i_2, j_2), \mu_2] \text{ iff } (i_1, j_1) \preceq_v (i_2, j_2) \text{ and } \mu_1 \preceq_d \mu_2.$$

There are two kinds of epistemic negation ($\neg_1$ and $\neg_2$) in EVALPSN, which are defined as mappings over the complete lattices $\mathcal{T}_v(n)$ and $\mathcal{T}_d$, respectively.

Definition 2.4 *(epistemic negations $\neg_1$ and $\neg_2$ in EVALPSN)*

$$\neg_1([(i, j), \mu]) = [(j, i), \mu], \quad \forall \mu \in \mathcal{T}_d$$
$$\neg_2([(i, j), \bot]) = [(i, j), \bot], \quad \neg_2([(i, j), \alpha]) = [(i, j), \alpha],$$
$$\neg_2([(i, j), \beta]) = [(i, j), \gamma], \quad \neg_2([(i, j), \gamma]) = [(i, j), \beta],$$

$$\neg_2([(i, j), *_1]) = [(i, j), *_3], \quad \neg_2([(i, j), *_2]) = [(i, j), *_2],$$
$$\neg_2([(i, j), *_3]) = [(i, j), *_1], \quad \neg_2([(i, j), \top]) = [(i, j), \top].$$

If we regard the epistemic negations in Definition 2.4 as syntactical operations, an epistemic negation followed by a literal can be eliminated by the syntactical operation. For example, $\neg_1 p : [(2, 0), \alpha] \leftrightarrow p : [(0, 2), \alpha]$ and $\neg_2 q : [(1, 0), \beta] \leftrightarrow p : [(1, 0), \gamma]$. The strong negation ($\sim$) in EVALPSN is defined as well as the paraconsistent annotated logic $P\mathcal{T}$.

Definition 2.5 (*well extended vector annotated literal*) Let p be a literal. $p : [(i, 0), \mu]$ and $p : [(0, j), \mu]$ are called *weva*(*well extended vector annotated*)-*literals*, where $i, j \in \{1, 2, \ldots, n\}$, and $\mu \in \{\alpha, \beta, \gamma\}$.

Definition 2.6 (*EVALPSN*) If $L_0, \ldots, L_n$ are weva-literals,

$$L_1 \wedge \cdots \wedge L_i \wedge \sim L_{i+1} \wedge \cdots \wedge \sim L_n \rightarrow L_0$$

is called an *EVALPSN clause*. An *EVALPSN* is a finite set of EVALPSN clauses.

Fact and deontic notions, "obligation", "forbiddance" and "permission" are represented by extended vector annotations,

$$[(m, 0), \alpha], \; [(m, 0), \beta], \; [(0, m), \beta], \; \text{and} \; [(0, m), \gamma],$$

respectively, where m is a positive integer. For example,

$p : [(2, 0), \alpha]$ is intuitively interpreted as "it is known to be true of strength 2 that p is a fact";

$p : [(1, 0), \beta]$ is as "it is known to be true of strength 1 that p is obligatory";

$p : [(0, 2), \beta]$ is as "it is known to be false of strength 2 that p is obligatory", that is to say, "it is known to be true of strength 2 that p is forbidden";

$p : [(0, 1), \gamma]$ is as "it is known to be false of strength 1 that p is not obligatory", that is to say, "it is known to be true of strength 1 that p is permitted".

Generally, if an EVALPSN contains the strong negation $\sim$, it has stable model semantics [39] as well as other ordinary logic programs with strong negation. However, the stable model semantics may have a problem that some programs may have more than two stable models and others have no stable model. In addition, computation of stable models takes a long time compared to usual logic programming such as PROLOG programming.

Therefore, it does not seem to be so appropriate for practical application such as real time processing in general. However, we fortunately have cases to implement EVALPSN practically, if an EVALPSN is a *stratified* program, it has a tractable model called a *perfect model* [46] and the strong negation in the EVALPSN can be treated as the *Negation as Failure* in logic programming with no strong negation.

The details of stratified program and some tractable models for normal logic programs can be found in [4, 12, 46, 48], furthermore the details of the stratified

EVALPSN are described in [39]. Therefore, inefficient EVALPSN stable model computation does not have to be taken into account in this chapter since all EVALPSNs that will appear in the subsequent sections are stratified.

7.3   Traffic Signal Control in EVALPSN

7.3.1   Deontic Defeasible Traffic Signal Control

It is clear that we should resolve traffic jam caused by inappropriate traffic signal control. In this section, we introduce an intelligent traffic signal control system based on EVALPSN defeasible deontic reasoning, which may provide one solution for traffic jam reduction. We show how the traffic signal control is implemented in EVALPSN with taking a simple intersection example in Japan. We suppose an intersection in which two roads are crossing described in Fig. 7.3 as an example for implementing the traffic signal control method based on EVALPSN.[2]

The intersection has four traffic lights $T_{1,2,3,4}$, which indicate four kinds of signals, green, yellow, red and right-turn arrow. Each lane connected to the intersection has a sensor to detect traffic amount. Each sensor is described by symbols S_i? ($1 \leq i \leq 8$) in Fig. 7.3.

For example, the sensor S_6 detects the right-turn traffic amount confronting traffic light T_1. Basically, the traffic signal control is performed based on the traffic amount detected by the sensors. The chain of signaling is supposed as follows:

$$\rightarrow \text{ red } \rightarrow \text{ green } \rightarrow \text{ yellow } \rightarrow \text{ right arrow } \rightarrow \text{ red } \rightarrow .$$

For simplicity, we assume that the durations of yellow and right arrow signals are constant, and if traffic lights $T_{1,2}(T_{3,4})$ are green or right arrow, traffic lights $T_{3,4}(T_{1,2})$ are red as follows:

Signal cycle of traffic lights $T_{1,2}$

$$\rightarrow \text{ red } \rightarrow \text{ red } \rightarrow \text{ green } \rightarrow \text{ right arrow } \rightarrow \text{ red } \rightarrow,$$

Signal cycle of traffic lights $T_{3,4}$

$$\rightarrow \text{ green } \rightarrow \text{ right arrow } \rightarrow \text{ red } \rightarrow \text{ red } \rightarrow \text{ green } \rightarrow .$$

Only the turns green to right arrow and right arrow to red are controlled. The turn red to green of the front traffic signal follows the turn right arrow to red of the neighbor one. Moreover, the signaling is controlled at each unit time $t \in \{0, 1, 2, \ldots, n\}$. The traffic amount of each lane can be regarded as permission or forbiddance from turning such as green to right arrow.

---

[2]The intersection is supposed to be in Japan where we need to keep left if driving a car.

Fig. 7.3 Intersection

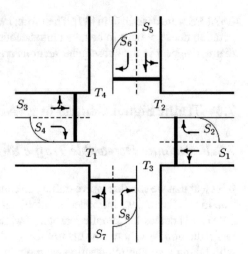

For example, if there are many cars waiting for traffic lights $T_{1,2}$ turning red to green, it can be regarded as permission for turning the crossing traffic lights $T_{3,4}$ green to right arrow, yellow and red. On the other hand, if there are many cars passing through the intersection with traffic lights $T_{3,4}$ signaling green, it can be regarded as forbiddance from turning traffic lights $T_{3,4}$ green to right arrow. Then, there is a conflict between those permission and forbiddance in terms of the same traffic lights $T_{3,4}$.

We formalize such a conflict resolving in EVALPSN. We assume that the minimum and maximum durations of green signal are previously given for all traffic lights, and the duration of green signal must be controlled between the minimum and maximum durations. We consider the following four states of traffic lights $T_{1,2,3,4}$,

state 1 traffic lights $T_{1,2}$ are red and traffic lights $T_{3,4}$ are green,
state 2 traffic lights $T_{1,2}$ are red and traffic lights $T_{3,4}$ are right arrow,
state 3 traffic lights $T_{1,2}$ are green and traffic lights $T_{3,4}$ are red,
state 4 traffic lights $T_{1,2}$ are right arrow and traffic lights $T_{3,4}$ are red.

Here we take the transit from the state 1 to the state 2 into account to introduce the traffic signal control properties in the state 1 and its translation into EVALPSN. The traffic signal control consists of the traffic signal control properties for the state transit the state 1 to the state 2, green light length rules, and deontic defeasible reasoning rules for traffic signal control.

We use the following EVALP literals:

- $S_i(t) : [(2, 0), \alpha]$ can be informally interpreted as the traffic sensor $S_i (i = 1, 2, \ldots, 8)$ has detected traffic at time t.
- $T_{m,n}(c, t) : [(2, 0), \alpha]$ can be informally interpreted as the traffic light $T_{m,n}$ indicates a signal color C at time t, where $m, n = 1, 2, 3, 4$ and c is one of signal colors green(g), red(r), or right arrow(a).

- $MIN_{m,n}(g,t):[(2,0),\alpha]$ can be informally interpreted as the green duration of traffic lights $T_{m,n}(m,n=1,2,3,4)$ is shorter than its minimum green duration at time t.
- $MAX_{m,n}(g,t):[(2,0),\alpha]$ can be informally interpreted as the green duration of traffic lights $T_{m,n}$ is longer than its maximum green duration at time t.
- $T_{m,n}(c,t):[(0,k),\gamma]$ which can be informally interpreted as it is permitted for traffic lights $T_{m,n}$ to indicate signal color C at time t, where $m,n=1,2,3,4$ and c is one of the signal colors green(g), red(r), or right arrow(a); if $k=1$, the permission is weak, and if $k=2$, the permission is strong.
- $T_{m,n}(c,t):[(0,k),\beta]$ can be informally interpreted as it is forbidden for traffic lights $T_{m,n}$ from indicating the signal color C at time t, where $m,n=1,2,3,4$ and c is one of the signal colors green(g), red(r), or right arrow(a); if $k=1$, the forbiddance is weak, and if $k=2$, the forbiddance is strong.

[Traffic Signal Control Properties in State 1]

1 If traffic sensor S_1 detects traffic amount, it has already passed the minimum green duration of traffic lights $T_{3,4}$, and neither traffic sensors S_5 nor S_7 detect traffic amount at time t, then it is weakly permitted for traffic lights $T_{3,4}$ to turn green to right arrow at time t; which is translated into the EVALPSN,

$$S_1(t):[(2,0),\alpha] \wedge$$
$$T_{1,2}(r,t):[(2,0),\alpha] \wedge T_{3,4}(g,t):[(2,0),\alpha] \wedge$$
$$\sim MIN_{3,4}(g,t):[(2,0),\alpha] \wedge$$
$$\sim S_5(t):[(2,0),\alpha] \wedge \sim S_7(t):[(2,0),\alpha]$$
$$\rightarrow T_{3,4}(a,t):[(0,1),\gamma]. \tag{7.1}$$

2 If traffic sensor S_3 detects traffic amount, it has already passed the minimum green duration of traffic lights $T_{3,4}$, and neither traffic sensors S_5 nor S_7 detect traffic amount at time t, then it is weakly permitted for traffic lights $T_{3,4}$ to turn green to right arrow at time t; which is translated into the EVALPSN,

$$S_3(t):[(2,0),\alpha] \wedge$$
$$T_{1,2}(r,t):[(2,0),\alpha] \wedge T_{3,4}(g,t):[(2,0),\alpha] \wedge$$
$$\sim MIN_{3,4}(g,t):[(2,0),\alpha] \wedge$$
$$\sim S_5(t):[(2,0),\alpha] \wedge \sim S_7(t):[(2,0),\alpha]$$
$$\rightarrow T_{3,4}(a,t):[(0,1),\gamma]. \tag{7.2}$$

3 If traffic sensor S_2 detects traffic amount, it has already passed the minimum green duration of traffic lights $T_{3,4}$, and neither traffic sensors S_5 nor S_7 detect traffic amount at time t, then it is weakly permitted for traffic lights $T_{3,4}$ to turn green to right arrow at time t, which is translated into the EVALPSN,

$$S_2(t):[(2,0),\alpha]\wedge$$
$$T_{1,2}(r,t):[(2,0),\alpha]\wedge T_{3,4}(g,t):[(2,0),\alpha]\wedge$$
$$\sim MIN_{3,4}(g,t):[(2,0),\alpha]\wedge$$
$$\sim S_5(t):[(2,0),\alpha]\wedge\sim S_7(t):[(2,0),\alpha]$$
$$\rightarrow T_{3,4}(a,t):[(0,1),\gamma], \tag{7.3}$$

4 If traffic sensor S_4 detects traffic amount, it has already passed the minimum green duration of traffic lights $T_{3,4}$, and neither traffic sensors S_5 nor S_7 detect traffic amount at time t, then it is weakly permitted for traffic lights $T_{3,4}$ to turn green to right arrow at time t; which is translated into the EVALPSN,

$$S_4(t):[(2,0),\alpha]\wedge$$
$$T_{1,2}(r,t):[(2,0),\alpha]\wedge T_{3,4}(g,t):[(2,0),\alpha]\wedge$$
$$\sim MIN_{3,4}(g,t):[(2,0),\alpha]\wedge$$
$$\sim S_5(t):[(2,0),\alpha]\wedge\sim S_7(t):[(2,0),\alpha]$$
$$\rightarrow T_{3,4}(a,t):[(0,1),\gamma], \tag{7.4}$$

5 If traffic sensor S_6 detects traffic amount, it has already passed the minimum green duration of traffic lights $T_{3,4}$, and neither traffic sensors S_5 nor S_7 detect traffic amount at time t, then it is weakly permitted for traffic lights $T_{3,4}$ to turn green to right arrow at time t; which is translated into the EVALPSN,

$$S_6(t):[(2,0),\alpha]\wedge$$
$$T_{1,2}(r,t):[(2,0),\alpha]\wedge T_{3,4}(g,t):[(2,0),\alpha]\wedge$$
$$\sim MIN_{3,4}(g,t):[(2,0),\alpha]\wedge$$
$$\sim S_5(t):[(2,0),\alpha]\wedge\sim S_7(t):[(2,0),\alpha]$$
$$\rightarrow T_{3,4}(a,t):[(0,1),\gamma], \tag{7.5}$$

6 If traffic sensor S_8 detects traffic amount, it has already passed the minimum green duration of traffic lights $T_{3,4}$, and neither traffic sensors S_5 nor S_7 detect traffic amount at time t, then it is weakly permitted for traffic lights $T_{3,4}$ to turn green to right arrow at time t; which is translated into the EVALPSN,

$$S_6(t):[(2,0),\alpha]\wedge$$
$$T_{1,2}(r,t):[(2,0),\alpha]\wedge T_{3,4}(g,t):[(2,0),\alpha]\wedge$$
$$\sim MIN_{3,4}(g,t):[(2,0),\alpha]\wedge$$
$$\sim S_5(t):[(2,0),\alpha]\wedge\sim S_7(t):[(2,0),\alpha]$$
$$\rightarrow T_{3,4}(a,t):[(0,1),\gamma], \tag{7.6}$$

7 If traffic sensor S_5 detects traffic amount and it has not passed the maximum green duration of traffic lights $T_{3,4}$ yet, then it is weakly forbidden for traffic lights $T_{3,4}$ to turn green to right arrow at time t; which is translated into the EVALPSN,

$$S_5(t):[(2,0),\alpha] \wedge$$
$$T_{1,2}(r,t):[(2,0),\alpha] \wedge T_{3,4}(g,t):[(2,0),\alpha] \wedge$$
$$\sim\!MAX_{3,4}(g,t):[(2,0),\alpha]$$
$$\rightarrow T_{3,4}(a,t):[(0,1),\beta], \tag{7.7}$$

8 If traffic sensor S_7 detects traffic amount and it has not passed the maximum green duration of traffic lights $T_{3,4}$, then it is weakly forbidden for traffic lights $T_{3,4}$ to turn green to right arrow at time t; which is translated into the EVALPSN,

$$S_7(t):[(2,0),\alpha] \wedge$$
$$T_{1,2}(r,t):[(2,0),\alpha] \wedge T_{3,4}(g,t):[(2,0),\alpha] \wedge$$
$$\sim\!MAX_{3,4}(g,t):[(2,0),\alpha]$$
$$\rightarrow T_{3,4}(a,t):[(0,1),\beta], \tag{7.8}$$

[Green light length rules for the traffic lights $T_{3,4}$]

9 If traffic lights $T_{3,4}$ are green and it has not passed the minimum duration of them yet, then it is strongly forbidden for traffic lights $T_{3,4}$ to turn green to right arrow at time t; which is translated into the EVALPSN,

$$T_{3,4}(g,t):[(2,0),\alpha] \wedge MIN_{3,4}(g,t):[(2,0),\alpha]$$
$$\rightarrow T_{3,4}(a,t):[(0,2),\beta], \tag{7.9}$$

10 If traffic lights $T_{3,4}$ are green and it has already passed the maximum duration of them, then it is strongly permitted for traffic lights $T_{3,4}$ to turn green to right arrow at time t; which is translated into the EVALPSN,

$$T_{3,4}(g,t):[(2,0),\alpha] \wedge MAX_{3,4}(g,t):[(2,0),\alpha]$$
$$\rightarrow T_{3,4}(a,t):[(0,2),\gamma], \tag{7.10}$$

[Deontic deasible reasoning rules]

11 If traffic lights $T_{3,4}$ are green, it is weakly permitted at least for traffic lights $T_{3,4}$ to turn green to right arrow at time t, then it is strongly obligatory for traffic lights $T_{3,4}$ to turn green to right arrow at time $t+1$ (at the next step); which is translated into the EVALPSN,

$$T_{3,4}(g,t):[(2,0),\alpha] \wedge T_{3,4}(a,t):[(0,1),\gamma]$$
$$\rightarrow T_{3,4}(a,t+1):[(2,0),\beta], \tag{7.11}$$

12 If traffic lights $T_{3,4}$ are green, it is weakly forbidden at least for traffic lights $T_{3,4}$ to turn green to right arrow at time t, then it is strongly obligatory for traffic lights $T_{3,4}$ not to turn green to right arrow at time $t + 1$ (at the next step); which is translated into the EVALPSN,

$$T_{3,4}(g, t) : [(2, 0), \alpha] \wedge T_{3,4}(a, t) : [(0, 1), \beta]$$
$$\rightarrow T_{3,4}(g, t + 1) : [(2, 0), \beta]. \tag{7.12}$$

7.3.2 Example and Simulation

Let us introduce a simple example of the EVALPSN based traffic signal control. We assume the same intersection in the previous section.

Example 3.1 Suppose that traffic lights $T_{1,2}$ are red and traffic lights $T_{3,4}$ are green. We also suppose that the minimum duration of green signal has already passed but the maximum one has not passed yet. Then, we obtain the EVALPSN,

$$T_{1,2}(r, t) : [(2, 0), \alpha] \wedge T_{3,4}(g, t) : [(2, 0), \alpha], \tag{7.13}$$

$$\sim MIN_{3,4}(g, t) : [(2, 0), \alpha], \tag{7.14}$$

$$\sim MAX_{3,4}(g, t) : [(2, 0), \alpha], \tag{7.15}$$

If traffic sensors $S_{1,3,5}$ detect traffic amount and traffic sensors $S_{2,4,6,7,8}$ do not detect traffic amount at time t, we obtain the EVALPSN,

$$S_1(t) : [(2, 0), \alpha], \tag{7.16}$$

$$S_3(t) : [(2, 0), \alpha], \tag{7.17}$$

$$S_5(t) : [(2, 0), \alpha], \tag{7.18}$$

$$\sim S_2(t) : [(2, 0), \alpha], \tag{7.19}$$

$$\sim S_4(t) : [(2, 0), \alpha], \tag{7.20}$$

$$\sim S_6(t) : [(2, 0), \alpha], \tag{7.21}$$

$$\sim S_7(t) : [(2, 0), \alpha], \tag{7.22}$$

$$\sim S_8(t) : [(2, 0), \alpha]. \tag{7.23}$$

Then, by EVALPSN clauses (7.7), (7.13), (7.15) and (7.18) the forbiddance from traffic lights $T_{3,4}$ turning to right arrow,

$$T_{3,4}(a, t) : [(0, 1), \beta] \tag{7.24}$$

is derived, furthermore, by EVALPSN clauses (7.12), (7.13) and (7.24) the obligation for traffic lights $T_{3,4}$ keeping green at time $t + 1$,

$$T_{3,4}(g, t + 1) : [(2, 0), \beta]$$

is obtained.

On the other hand, if traffic sensors $S_{1,3}$ detect traffic amount and traffic sensors $S_{2,4,5,6,7,8}$ do not detect traffic amount at time t, we obtain the EVALPSN,

$$S_1(t) : [(2, 0), \alpha], \tag{7.25}$$
$$S_3(t) : [(2, 0), \alpha], \tag{7.26}$$
$$\sim S_2(t) : [(2, 0), \alpha], \tag{7.27}$$
$$\sim S_4(t) : [(2, 0), \alpha], \tag{7.28}$$
$$\sim S_5(t) : [(2, 0), \alpha], \tag{7.29}$$
$$\sim S_6(t) : [(2, 0), \alpha], \tag{7.30}$$
$$\sim S_7(t) : [(2, 0), \alpha], \tag{7.31}$$
$$\sim S_8(t) : [(2, 0), \alpha]. \tag{7.32}$$

Then, by EVALPSN clauses (7.1), (7.13), (7.14), (7.25), (7.29) and (7.31), the permission for traffic lights $T_{3,4}$ turning to right arrow,

$$T_{3,4}(a, t) : [(0, 1), \gamma] \tag{7.33}$$

is derived, furthermore, by EVALPSN clauses (7.11), (7.13) and (7.33) the obligation for traffic lights $T_{3,4}$ turning to right arrow at time $t + 1$,

$$T_{3,4}(a, t + 1) : [(2, 0), \beta]$$

is finally obtained.

Here we introduce an EVALPSN traffic control simulation system based on the cellular automaton method and its simulation results comparing to ordinary fixed-time traffic signal control. In order to evaluate the simulation results we define the concepts "step", "move times", and "stop times" as follows:

step a time unit in the simulation system, which is a transit time that one car moves from its current cell to the next cell.

move times shows the times that one car moves from its current cell to the next cell without stop.

stop times shows the times that one car stops during transition from one cell to another cell.

Table 7.1 Simulation results

	Fixed-time control		EVALPSN control	
	Stop times	Move times	Stop times	Move times
Condition 1	17690	19641	16285	23151
Condition 2	16764	18664	12738	20121

We introduce the simulation results under the following two traffic flow conditions.

[Condition 1]

Cars are supposed to flow into the intersection from each road with the same probabilities, right-turn 5 %, left-turn 5 % and straight 20 %. It is supposed that green signal duration is 30 steps, yellow one is 3 steps, right-arrow one is 4 steps and red one is 40 steps in the fixed-time traffic signal control. It is also supposed that green signal duration is between 14 and 30 steps in the EVALPSN traffic signal control.

[Condition 2]

Cars are supposed to flow into the intersection with the following probabilities,

from South	right-turn 5 %, left-turn 15 % and straight 10 %;
from North	right-turn 15 %, left-turn 5 % and straight 10 %;
from West	right-turn, left-turn and straight 5 % each;
from East	right-turn and left-turn 5 % each, and straight 15 %.

Other conditions are the same as the Condition 1.

We measured the numbers of car stop and move times during 1000 steps, and repeated it 10 times under the same conditions. The average numbers of car stop and move times are listed in Table 7.1. The simulation results show that the number of car move times in the EVALPSN traffic signal control is larger than that in the fixed-time traffic signal control, and the number of car stop times in the EVALPSN traffic signal control is smaller than that in the fixed time one. Taking the simulation results into account, it could be concluded that the EVALPSN traffic signal control is more efficient for relieving traffic congestion than the fixed-time traffic signal control.

7.4 EVALPSN Safety Verification for Pipeline Control

This section introduces EVALPSN based safety verification for pipeline valve control with a simple brewery pipeline example.

7.4.1   Pipeline Network

The pipeline network described in Fig. 7.4 is taken as an example for the brewery pipeline valve control based on EVALPSN safety verification. In Fig. 7.4, the arrows represent the directions of liquid flows, home-plate pentagons show brewery tanks, and cross figures do valves.

In the pipeline network, we suppose physical entities:

- four tanks, T_0, T_1, T_2, and T_3;
- five pipes, Pi_0, Pi_1, Pi_2, Pi_3, and Pi_4;
  (a pipe includes neither valves nor tanks)
- two valves, V_0, and V_1;

and logical entities that we suppose:

- four processes, Pr_0, Pr_1, Pr_2, and Pr_3;
  (a process is defined as a set of sub-processes and valves)
- five sub-processes, SPr_0, SPr_1, SPr_2, SPr_3, and SPr_4.

For example, process Pr_0 consists of sub-processes SPr_0, SPr_1 and valve V_0. Each entity is supposed to have logical or physical states. Sub-processes have two states *locked*(1) and *free*(f), then "the sub-process is locked" means that the sub-process is supposed to be locked(logically reserved) by one sort of liquid, and "free" means unlocked. Processes have two states *set*(s) and *unset*(xs), then "the process is set" means that all the sub-processes in the process are locked, and "unset" means not set.

Here we also assume that valves in the network can control two liquid flows in the normal and cross directions as shown in Fig. 7.5.

Valves have two controlled states, *controlled mix*(cm) representing that the valve is controlled to mix the liquid flows in the normal and cross directions, and *controlled*

Fig. 7.4 Pipeline example

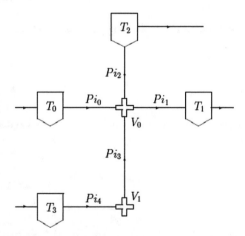

Fig. 7.5 Normal and cross
directions

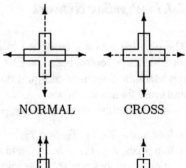

NORMAL   CROSS

Fig. 7.6 Controlled mix and
separate

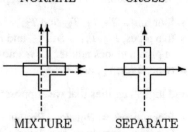

MIXTURE   SEPARATE

separate(cs) representing that the valve is controlled to separate the liquid flow in
the normal and cross directions as shown in Fig. 7.6.

We suppose that there are five sorts of cleaning liquid:

cold water(*cw*), warm water(*ww*), hot water(*hw*),

nitric acid(*na*) and caustic soda(*cs*).

We also consider the following four brewery and cleaning processes in the pipeline
network:

- Process Pr_0 a beer process,

$$T_0 \longrightarrow V_0(\text{cs}) \longrightarrow T_1$$

- Process Pr_1 a cleaning process with nitric acid,

$$T_2$$

$$\uparrow$$

$$V_0(\text{cs})$$

$$\uparrow$$

$$T_3 \longrightarrow V_1(\text{cm})$$

- Process Pr_2 a cleaning process with cold water,

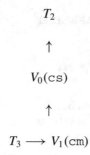

$$T_2$$

$$\uparrow$$

$$V_0(\text{cs})$$

$$\uparrow$$

$$T_3 \longrightarrow V_1(\text{cm})$$

- Process Pr_3 a brewery mixing process,

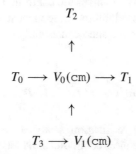

$$T_2$$

$$\uparrow$$

$$T_0 \longrightarrow V_0(\text{cm}) \longrightarrow T_1$$

$$\uparrow$$

$$T_3 \longrightarrow V_1(\text{cm})$$

In order to verify the safety for the above processes, the pipeline controller issues a process request consisting of if-part and then-part before starting the process. The if-part describes the current state of the pipelines that should be used in the process, and the then-part describes the permission for setting the process. For example, the process request for process Pr_1 is described as:

>   if   sub-process SPr_0 is free,
>   sub-process SPr_1 is free,
>   valve V_0 is physically controlled separate and free,
>   then   process Pr_0 can be set ?

We also suppose the following process schedule for processes $Pr_{0,1,2,3}$:

- S-0 process Pr_0 starts before any other processes;
- S-1 process Pr_1 starts immediately after process Pr_0;
- S-2 process Pr_2 starts immediately after process Pr_1;
- PRS-3 process Pr_3 starts immediately after processes Pr_0 and Pr_2,

which are charted in Fig. 7.7.

Fig. 7.7 Process Schedule
Chart

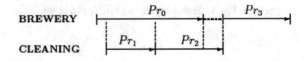

7.4.2 Pipeline Safety Property

We introduce the safety properties, **SPr** for sub-processes, **Val** for valves, and **Pr** for processes, for assuring the pipeline valve control safety, which can avoid unexpected mix of different sorts of liquid in the valve.

- **SPr**: it is a forbidden case that the sub-process over a given pipe is simultaneously locked by different sorts of liquid.
- **Val**: it is a forbidden case that valves are controlled for unexpected mix of liquid.
- **Pr**: whenever a process is set, all its component sub-processes are locked and all its component valves are controlled consistently.

7.4.3 Predicates for Safety Verification

The EVALPSN based safety verification is carried out by verifying whether process start requests by pipeline operators contradict the safety properties or not in EVALPSN programming. Then the following three steps 1, 2 and 3 have to be executed:

1. the safety properties for the pipeline network, which should be insured when the pipeline network is locked, and some control methods for the network are translated into EVALPSN clauses, and they have to be stored as EVALPSN P_{sc};
2. the if-part of the process request that is the current state of the pipeline and the then-part of the process request that is supposed to be verified are translated into EVALP clauses as EVALPs P_i and P_t, respectively;
3. EVALP P_t is inquired from EVALPSN $\{P_{sc} \cup P_i\}$, then if *yes* is returned, the safety for the request is assured and the defeasible deontic reasoning is performed, otherwise, it is not assured; which is described in Fig. 7.8.

In order to verify the safety for the pipeline network, the following predicates are used in EVALPSN.

- $Pr(i, l)$ represents that the process i for the liquid l is set(s) or unset(xs), where

$$i \in \{p0, p1, p2, p3\}$$

is a process id corresponding to processes $Pr_{0,1,2,3}$, respectively,

$$l \in \{b, cw, ww, hw, na, cs\}$$

is a liquid sort, and we have the EVALP clause,

Fig. 7.8 EVALPSN based
safety verification

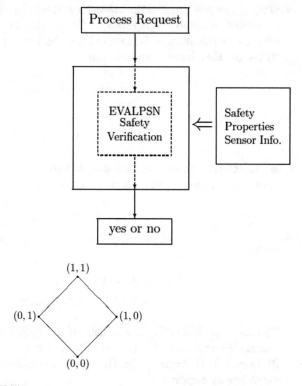

Fig. 7.9 The complete lattice $\mathcal{T}_v(1)$

$$Pr(i, l):[\mu_1, \mu_2],$$

where

$$\mu_1 \in \mathcal{T}_{v_1} = \{\perp_1, s, xs, \top_1\},$$
$$\mu_2 \in \mathcal{T}_d = \{\perp, \alpha, \beta, \gamma, *_1, *_2, *_3, \top\}.$$

The complete lattice $\mathcal{T}_{v_1}$ is a variant of the complete lattice $\mathcal{T}_v(1)$ in Fig. 7.9. Therefore annotations $\perp_1, s, xs$ and $\top_1$ are for vector annotations $(0, 0), (1, 0), (0, 1)$ and $(1, 1)$, respectively.

The epistemic negation $\neg_1$ over $\mathcal{T}_{v_1}$ is defined as the following mapping:

$$\neg_1([\perp_1, \mu_2]) = [\perp_1, \mu_2], \quad \neg_1([s, \mu_2]) = [xs, \mu_2],$$
$$\neg_1([\top_1, \mu_2]) = [\top_1, \mu_2], \quad \neg_1([xs, \mu_2]) = [s, \mu_2].$$

For example, EVALP clause $Pr(p2, b):[s, \alpha]$ can be intuitively interpreted as "it is a fact that the beer process Pr_2 is set".

- $SPr(i, j, l)$ represents that the sub-process from valve(or tank) i to valve(or tank) j occupied by liquid l is locked(1) or free(f). Moreover, if a sub-process is free, the liquid sort in the pipe is not cared, and the liquid is represented by the symbol "0"(zero). Therefore we suppose that

$$l \in \{b, cw, ww, hw, na, cs, 0\}$$

and

$$i, j \in \{v0, v1, t0, t1, t2, t3\}$$

are valve and tank ids corresponding to valves $V_{0,1}$, and tanks $T_{0,1,2,3}$. Then we have the following EVALP clause for representing sub-process states:

$$SPr(i, j, l) : [\mu_1, \mu_2],$$

where

$$\mu_1 \in \mathcal{T}_{v_2} = \{\bot_2, 1, f, \top_2\},$$
$$\mu_2 \in \mathcal{T}_d = \{\bot, \alpha, \beta, \gamma, *_1, *_2, *_3, \top\}.$$

The complete lattice $\mathcal{T}_{v_2}$ is a variant of the complete lattice $\mathcal{T}_v(1)$ in Fig. 7.9. Therefore annotations $\bot_2, 1, f$ and $\top_2$ are for vector annotations $(0, 0)$, $(1, 0)$, $(0, 1)$ and $(1, 1)$, respectively. The epistemic negation $\neg_1$ over $\mathcal{T}_{v_2}$ is defined as the following mapping:

$$\neg_1([\bot_2, \mu_2]) = [\bot_2, \mu_2], \quad \neg_1([1, \mu_2]) = [f, \mu_2],$$
$$\neg_1([\top_2, \mu_2]) = [\top_2, \mu_2], \quad \neg_1([f, \mu_2]) = [1, \mu_2].$$

For example, EVALP clause $SPr(v0, t1, b) : [f, \gamma]$ can be intuitively interpreted as "the sub-process from valve V_0 to tank T_1 is permitted to be locked by the beer process".

- $Val(i, l_n, l_c)$ represents that valve i occupied by the two sorts of liquid $l_n, l_c \in \{b, cw, ww, hw, na, cs, 0\}$ is controlled separate(cs) or mix(cm), where $i \in \{v0, v1\}$ is a valve id. We suppose that there are two directed liquid flows in the *normal* and *cross* directions in valves as shown in Fig. 7.5. Therefore, the second argument l_n represents the liquid flowing in the normal direction and the third argument l_c represents the liquid flowing in the cross direction. Generally, if a valve is released from the locked(controlled) state, the liquid flow in the valve is represented by the symbol 0 that means "free". Then we have the following EVALP clause for representing valve states,

$$Val(i, l_n, l_c) : [\mu_1, \mu_2],$$

where

$$\mu_1 \in \mathcal{T}_{v_3} = \{\bot_3, \text{cm}, \text{cs}, \top_3\},$$
$$\mu_2 \in \mathcal{T}_d = \{\bot, \alpha, \beta, \gamma, *_1, *_2, *_3, \top\},$$

The complete lattice $\mathcal{T}_{v_3}$ is a variant of the complete lattice $\mathcal{T}_v(1)$ in Fig. 7.9. Therefore annotations $\bot_3$, cm, cs and $\top_3$ are for vector annotations $(0, 0)$, $(1, 0)$, $(0, 1)$ and $(1, 1)$, respectively. The epistemic negation $\neg_1$ over $\mathcal{T}_{v_3}$ is defined as the following mapping:

$$\neg_1([\bot_3, \mu_2]) = [\bot_3, \mu_2], \quad \neg_1([\text{cs}, \mu_2]) = [\text{cm}, \mu_2],$$
$$\neg_1([\top_3, \mu_2]) = [\top_3, \mu_2], \quad \neg_1([\text{cm}, \mu_2]) = [\text{cs}, \mu_2].$$

We suppose that if a process finishes, all valves included in the process are controlled separate(closed). For example, EVALP clause $Val(v0, 0, 0):[\text{cs}, \alpha]$ can be intuitively interpreted as "valve V_0 has been released from controlled separate state"; EVALP clause $Val(v0, b, cw):[\text{cs}, \beta]$ can be intuitively interpreted as both "it is forbidden for valve V_0 is controlled mix with beer b in the normal direction and cold water cw in the cross direction", and "it is obligatory for valve V_0 to be controlled separate with beer b in the normal direction and cold water cw in the cross direction"; and EVALP clause $Val(v0, 0, b):[\text{cs}, \alpha]$ can be intuitively interpreted as "it is a fact that valve V_0 is controlled separate with the free flow 0 in the normal direction and beer b in the cross direction".

- $Eql(l_1, l_2)$ represents that liquids l_1 and l_2 are the same(sa) or different(di), where

$$l_1, l_2 \in \{b, cw, ww, hw, na, cs, 0\}.$$

We have the following EVALP clause for distinguishing liquid:

$$Eql(l_1, l_2):[\mu_1, \mu_2],$$

where

$$\mu_1 \in \mathcal{T}_{v_4} = \{\bot_4, \text{sa}, \text{di}, \top_4\},$$
$$\mu_2 \in \mathcal{T}_d = \{\bot, \alpha, \beta, \gamma, *_1, *_2, *_3, \top\}.$$

The complete lattice $\mathcal{T}_{v_4}$ is a variant of the complete lattice $\mathcal{T}_v(1)$ in Fig. 7.9. Therefore annotations $\bot_4$, sa, di and $\top_4$ are for vector annotations $(0, 0)$, $(1, 0)$, $(0, 1)$ and $(1, 1)$, respectively. The epistemic negation $\neg_1$ is defined as the following mapping:

$$\neg_1([\bot_4, \mu_2]) = [\bot_4, \mu_2], \quad \neg_1([\text{di}, \mu_2]) = [\text{sa}, \mu_2],$$
$$\neg_1([\top_4, \mu_2]) = [\top_4, \mu_2], \quad \neg_1([\text{sa}, \mu_2]) = [\text{di}, \mu_2].$$

Now we consider process release conditions when processes have finished and define some more predicates to represent the conditions. We suppose that if the terminal tank T_i of process Pr_j is filled with one sort of liquid, the finish signal $Fin(pj)$ of process Pr_j is issued.

- $Tan(ti, l)$ represents that tank T_i has been filled fully(fu) with liquid l or empty(em). Then we have the following EVALP clause for representing tank states:

$$Tan(ti, l):[\mu_1, \mu_2],$$

where $i \in \{0, 1, 2, 3\}$ $l \in \{b, cw, ww, hw, na, cs, 0\}$,

$$\mu_1 \in \mathcal{T}_{v_5} = \{\perp_5, \text{fu}, \text{em}, \top_5\},$$
$$\mu_2 \in \mathcal{T}_d = \{\perp, \alpha, \beta, \gamma, *_1, *_2, *_3, \top\}.$$

The complete lattice $\mathcal{T}_{v_5}$ is a variant of the complete lattice $\mathcal{T}_v(1)$ in Fig. 7.9. Therefore annotations $\perp_5$, fu, em and $\top_5$ are for vector annotations $(0, 0), (1, 0), (0, 1)$ and $(1, 1)$, respectively. The epistemic negation $\neg_1$ over $\mathcal{T}_{v_5}$ is defined as the following mapping:

$$\neg_1([\perp_5, \mu_2]) = [\perp_5, \mu_2], \quad \neg_1([\text{fu}, \mu_2]) = [\text{em}, \mu_2],$$
$$\neg_1([\top_5, \mu_2]) = [\top_5, \mu_2], \quad \neg_1([\text{em}, \mu_2]) = [\text{fu}, \mu_2].$$

Note that annotation $\perp_5$ can be intuitively interpreted to represent "filled with some amount of liquid but not fully", that is to say, "no information in terms of fullness". For example, EVALP clause $Tan(t2, 0):[\text{em}, \alpha]$ can be interpreted as "it is a fact that tank T_2 is empty".

- $Str(pi)$ represents that the start signal for process Pr_i is issued (is) or not (ni).
- $Fin(pj)$ represents that the finish signal for process Pr_j has been issued (is) or not (ni). Then we have the following EVALP clauses for representing start/finish information:

$$Str(pi):[\mu_1, \mu_2], \quad Fin(pi):[\mu_1, \mu_2],$$

where $i, j \in \{0, 1, 2, 3\}$,

$$\mu_1 \in \mathcal{T}_{v_6} = \{\perp_6, \text{ni}, \text{is}, \top_6\},$$
$$\mu_2 \in \mathcal{T}_d = \{\perp, \alpha, \beta, \gamma, *_1, *_2, *_3, \top\}.$$

The complete lattice $\mathcal{T}_{v_6}$ is a variant of the complete lattice $\mathcal{T}_v(1)$ in Fig. 7.9. Therefore annotations $\perp_6$, is, ni and $\top_6$ are for vector annotations $(0, 0), (1, 0), (0, 1)$ and $(1, 1)$, respectively. The epistemic negation $\neg_1$ over $\mathcal{T}_{v_6}$ is defined as the following mapping:

$$\neg_1([\perp_6, \mu_2]) = [\perp_6, \mu_2], \quad \neg_1([\text{is}, \mu_2]) = [\text{ni}, \mu_2],$$
$$\neg_1([\top_6, \mu_2]) = [\top_6, \mu_2], \quad \neg_1([\text{ni}, \mu_2]) = [\text{is}, \mu_2].$$

For example, EVALP clause $Fin(p3):$ [ni, α] can be interpreted as "it is a fact that the finish signal for process Pr_3 has not been issued yet".

7.4.4 Safety Property in EVALPSN

Here, we provide the formalization of all safety properties **SPr**, **Val** and **Pr** in EVALPSN.

SPr can be intuitively interpreted as derivation rules of forbiddance. If a sub-process from valve(or tank) i to valve(or tank) j is locked by one sort of liquid, it is forbidden for the sub-process to be locked by different sorts of liquid simultaneously. Thus, we have the following EVALPSN clause for representing such forbiddance for sub-processes:

$$SPr(i, j, l_1):[1, \alpha] \wedge \sim Eql(l_1, l_2):[\text{sa}, \alpha]$$
$$\rightarrow SPr(i, j, l_2):[\text{f}, \beta], \tag{7.34}$$

where $l_1, l_2 \in \{b, cw, ww, hw, na, cs\}$. Moreover, in order to derive permission for locking sub-processes we need the following EVALPSN clause:

$$\sim SPr(i, j, l):[\text{f}, \beta] \rightarrow SPr(i, j, l):[\text{f}, \gamma], \tag{7.35}$$

where $l \in \{b, cw, ww, hw, na, cs\}$.

Val also can be intuitively interpreted as derivation rules of forbiddance. We have to consider two cases: one is for deriving the forbiddance from changing the control state of the valve, and another one is for deriving the forbiddance from mixing different sorts of liquid without changing the control state of the valve.

Case 1

If a valve is controlled separate, it is forbidden for the valve to be controlled mix, conversely, if a valve is controlled mixture, it is forbidden for the valve to be controlled separate. Thus, generally we have the following EVALPSN clauses:

$$Val(i, l_n, l_c):[\text{cs}, \alpha] \wedge \sim Eql(l_n, 0):[\text{sa}, \alpha] \wedge$$
$$\sim Eql(l_c, 0):[\text{sa}, \alpha] \rightarrow Val(i, l_n, l_c):[\text{cs}, \beta], \tag{7.36}$$

$$Val(i, l_n, l_c):[\text{cm}, \alpha] \wedge \sim Eql(l_n, 0):[\text{sa}, \alpha] \wedge$$
$$\sim Eql(l_c, 0):[\text{sa}, \alpha] \rightarrow Val(i, l_n, l_c):[\text{cm}, \beta], \tag{7.37}$$

where $l_n, l_c \in \{b, cw, ww, hw, na, cs, 0\}$.

Case 2

In this case, we consider another forbiddance derivation case in which different sorts of liquid are mixed even if the valve control state is not changed. We have the following EVALPSN clauses:

$$Val(i, l_{n_1}, l_{c_1}):[\text{cs}, \alpha] \wedge \sim Eql(l_{n_1}, l_{n_2}):[\text{sa}, \alpha] \wedge$$
$$\sim Eql(l_{n_1}, 0):[\text{sa}, \alpha] \rightarrow Val(i, l_{n_2}, l_{c_2}):[\text{cm}, \beta], \tag{7.38}$$

$$Val(i, l_{n_1}, l_{c_1}):[\text{cs}, \alpha] \wedge \sim Eql(l_{c_1}, l_{c_2}):[\text{sa}, \alpha] \wedge$$
$$\sim Eql(l_{c_1}, 0):[\text{sa}, \alpha] \rightarrow Val(i, l_{n_2}, l_{c_2}):[\text{cm}, \beta], \tag{7.39}$$

$$Val(i, l_{n_1}, l_{c_1}):[\text{cm}, \alpha] \wedge \sim Eql(l_{n_1}, l_{n_2}):[\text{sa}, \alpha]$$
$$\rightarrow Val(i, l_{n_2}, l_{c_2}):[\text{cs}, \beta], \tag{7.40}$$

$$Val(i, l_{n_1}, l_{c_1}):[\text{cm}, \alpha] \wedge \sim Eql(l_{c_1}, l_{c_2}):[\text{sa}, \alpha]$$
$$\rightarrow Val(i, l_{n_2}, l_{c_2}):[\text{cs}, \beta], \tag{7.41}$$

where $l_{n_1}, l_{c_1} \in \{b, cw, ww, hw, na, cs, 0\}$ and $l_{n_2}, l_{c_2} \in \{b, cw, ww, hw, na, cs\}$.

Here note that EVALPSN clause $\sim Eql(l_n, 0):[\text{sa}, \alpha]$ shows that there does not exist information such that the normal direction with liquid l_n in the valve is free (not controlled). As well as the case of sub-processes, in order to derive permission for controlling valves, we need the following EVALPSN clauses:

$$\sim Val(i, l_n, l_c):[\text{cm}, \beta] \rightarrow Val(i, l_n, l_c):[\text{cm}, \gamma], \tag{7.42}$$

$$\sim Val(i, l_n, l_c):[\text{cs}, \beta] \rightarrow Val(i, l_n, l_c):[\text{cs}, \gamma], \tag{7.43}$$

where $l_n, l_c \in \{b, cw, ww, hw, na, cs, 0\}$.

Pr can be intuitively interpreted as derivation rules of permission and directly translated into EVALPSN clauses as if-then rules "if all the components of the process can be locked or controlled consistently, then the process can be set". For example, if the beer process Pr_0 consists of the sub-process from tank T_0 to valve V_0, valve V_0 with controlled separate by beer in the normal direction, and the sub-process from valve V_0 to tank T_1, then we have the following EVALP clause to derive the permission for setting process Pr_0.

Process Pr_0

$$SPr(t0, v0, b) : [\texttt{f}, \gamma] \wedge SPr(v0, t1, b) : [\texttt{f}, \gamma] \wedge$$
$$Val(v0, b, l) : [\texttt{cm}, \gamma] \wedge Tan(t0, b) : [\texttt{fu}, \alpha] \wedge$$
$$Tan(t1, 0) : [\texttt{em}, \alpha] \rightarrow Pr(p0, b) : [\texttt{xs}, \gamma]. \tag{7.44}$$

We also have the following EVALP clauses for setting the other processes.

Process Pr_1

$$SPr(t3, v1, na) : [\texttt{f}, \gamma] \wedge SPr(v1, v0, na) : [\texttt{f}, \gamma] \wedge$$
$$SPr(v0, t2, na) : [\texttt{f}, \gamma] \wedge Val(v0, l, na) : [\texttt{cm}, \gamma] \wedge$$
$$Val(v1, na, 0) : [\texttt{cs}, \gamma] \wedge Tan(t3, na) : [\texttt{fu}, \alpha] \wedge$$
$$Tan(t2, 0) : [\texttt{em}, \alpha] \rightarrow Pr(p1, na) : [\texttt{xs}, \gamma]. \tag{7.45}$$

Process Pr_2

$$SPr(t3, v1, cw) : [\texttt{f}, \gamma] \wedge SPr(v1, v0, cw) : [\texttt{f}, \gamma] \wedge$$
$$SPr(v0, t2, cw) : [\texttt{f}, \gamma] \wedge Val(v0, l, cw) : [\texttt{cm}, \gamma] \wedge$$
$$Val(v1, cw, 0) : [\texttt{cs}, \gamma] \wedge Tan(t3, cw) : [\texttt{fu}, \alpha] \wedge$$
$$Tan(t2, 0) : [\texttt{em}, \alpha] \rightarrow Pr(p2, cw) : [\texttt{xs}, \gamma]. \tag{7.46}$$

Process Pr_3

$$SPr(t0, v0, b) : [\texttt{f}, \gamma] \wedge SPr(t3, v1, b) : [\texttt{f}, \gamma] \wedge$$
$$SPr(v0, t1, b) : [\texttt{f}, \gamma] \wedge SPr(v0, t2, b) : [\texttt{f}, \gamma] \wedge$$
$$SPr(v1, v0, b) : [\texttt{f}, \gamma] \wedge Val(v0, b, b) : [\texttt{cs}, \gamma] \wedge$$
$$Val(v1, b, 0) : [\texttt{cs}, \gamma] \wedge Tan(t0, b) : [\texttt{fu}, \alpha] \wedge$$
$$Tan(t1, 0) : [\texttt{em}, \alpha] \wedge Tan(t3, b) : [\texttt{fu}, \alpha] \wedge$$
$$Tan(t2, 0) : [\texttt{em}, \alpha] \rightarrow Pr(p3, b) : [\texttt{xs}, \gamma]. \tag{7.47}$$

We suppose that $l \in \{b, cw, ww, hw, na, cs, 0\}$ in the above safety verification EVALPSN clauses for processes $Pr_{0,1,2,3}$.

7.4.5 Process Release Control in EVALPSN

In this subsection, we consider conditions for releasing process lock after the process has finished. For example, a process release condition can be expressed by "liquid l in tank T_j has been transferred into tank T_k in process Pr_i after process Pr_i has started, and the finish signal $Fin(pi)$ for process Pr_i is obtained". If the above condition is satisfied, the locked process Pr_i is allowed to be unset, and each component of the process is also allowed to be free. The release conditions for processes $Pr_{0,1,2,3}$ are formalized in the following EVALPSN clauses.

Process Pr_0

$$Str(p0):[\text{is},\alpha] \wedge Tan(t0,b):[\text{em},\alpha] \wedge$$
$$Tan(t1,b):[\text{fu},\alpha] \wedge Fin(p0):[\text{is},\alpha] \rightarrow$$
$$Pr(p0,b):[\text{s},\gamma], \tag{7.48}$$
$$Pr(p0,b):[\text{s},\gamma] \rightarrow SPr(t0,v0,0):[\text{l},\gamma], \tag{7.49}$$
$$Pr(p0,b):[\text{s},\gamma] \rightarrow SPr(v0,t1,0):[\text{l},\gamma], \tag{7.50}$$
$$Pr(p0,b):[\text{s},\gamma] \rightarrow Val(v0,0,l):[\text{cm},\gamma]. \tag{7.51}$$

Process Pr_1

$$Str(p1):[\text{is},\alpha] \wedge Tan(t3,na):[\text{em},\alpha] \wedge$$
$$Tan(t2,na):[\text{fu},\alpha] \wedge Fin(p1):[\text{is},\alpha] \rightarrow$$
$$Pr(p1,na):[\text{s},\gamma], \tag{7.52}$$
$$Pr(p1,na):[\text{s},\gamma] \rightarrow SPr(v0,t2,0):[\text{l},\gamma], \tag{7.53}$$
$$Pr(p1,na):[\text{s},\gamma] \rightarrow SPr(v1,v0,0):[\text{l},\gamma], \tag{7.54}$$
$$Pr(p1,na):[\text{s},\gamma] \rightarrow SPr(t3,v1,0):[\text{l},\gamma], \tag{7.55}$$
$$Pr(p1,na):[\text{s},\gamma] \rightarrow Val(v0,l,0):[\text{cm},\gamma], \tag{7.56}$$
$$Pr(p1,na):[\text{s},\gamma] \rightarrow Val(v1,0,0):[\text{cm}\gamma]. \tag{7.57}$$

Process Pr_2

$$Str(p2):[\text{is},\alpha] \wedge Tan(t3,cw):[\text{em},\alpha] \wedge$$
$$Tan(t2,cw):[\text{fu},\alpha] \wedge Fin(p2):[\text{is},\alpha] \rightarrow$$
$$Pr(p2,cw):[\text{s},\gamma], \tag{7.58}$$
$$Pr(p2,cw):[\text{s},\gamma] \rightarrow SPr(v0,t2,0):[\text{l},\gamma], \tag{7.59}$$
$$Pr(p2,cw):[\text{s},\gamma] \rightarrow SPr(v1,v0,0):[\text{l},\gamma], \tag{7.60}$$
$$Pr(p2,cw):[\text{s},\gamma] \rightarrow SPr(t3,v1,0):[\text{l},\gamma], \tag{7.61}$$
$$Pr(p2,cw):[\text{s},\gamma] \rightarrow Val(v0,l,0):[\text{cm},\gamma], \tag{7.62}$$
$$Pr(p2,cw):[\text{s},\gamma] \rightarrow Val(v1,0,0):[\text{cm},\gamma]. \tag{7.63}$$

Process Pr_3

$$Str(p3):[\text{is},\alpha] \wedge Tan(t0,b):[\text{em},\alpha] \wedge$$
$$Tan(t3,b):[\text{em},\alpha] \wedge Tan(t1,b):[\text{fu},\alpha] \wedge$$
$$Tan(t2,b):[\text{fu},\alpha] \wedge Fin(p3):[\text{is},\alpha] \rightarrow$$
$$Pr(p3,b):[\text{s},\gamma], \tag{7.64}$$

$$Pr(p3, b):[s, \gamma] \rightarrow SPr(t0, v0, 0):[1, \gamma], \tag{7.65}$$

$$Pr(p3, b):[s, \gamma] \rightarrow SPr(v0, t1, 0):[1, \gamma], \tag{7.66}$$

$$Pr(p3, b):[s, \gamma] \rightarrow SPr(v0, t2, 0):[1, \gamma], \tag{7.67}$$

$$Pr(p3, b):[s, \gamma] \rightarrow SPr(v1, v0, 0):[1, \gamma], \tag{7.68}$$

$$Pr(p3, b):[s, \gamma] \rightarrow SPr(t3, v1, 0):[1, \gamma], \tag{7.69}$$

$$Pr(p3, b):[s, \gamma] \rightarrow Val(v0, l, 0):[\text{cm}, \gamma], \tag{7.70}$$

$$Pr(p3, b):[s, \gamma] \rightarrow Val(v1, 0, 0):[\text{cm}, \gamma]. \tag{7.71}$$

We suppose that $l \in \{b, cw, ww, hw, na, cs, 0\}$ in the above process release EVALPSN clauses for processes $Pr_{0,1,2,3}$.

7.4.6 Example

In this subsection we introduce an example of EVALPSN safety verification based pipeline control for processes $Pr_{0,1,2,3}$ in the pipeline network in Fig. 7.4. According to the process schedule in Fig. 7.7, we describe the details of EVALPSN safety verification.

Initial Stage We suppose that all the sub-processes and valves in the pipeline network are free (unlocked) and no process has already started at the initial stage. In order to verify the safety for all processes $Pr_{0,1,2,3}$, the following fact EVALP clauses (detected information) are input to the pipeline safety control EVALPSN:

$$SPr(t0, v0, 0):[\text{f}, \alpha], \quad Val(v0, 0, 0):[\text{cs}, \alpha],$$

$$SPr(v0, t1, 0):[\text{f}, \alpha], \quad Val(v1, 0, 0):[\text{cs}, \alpha],$$

$$SPr(v0, t2, 0):[\text{f}, \alpha],$$

$$SPr(v1, v0, 0):[\text{f}, \alpha],$$

$$SPr(t3, v1, 0):[\text{f}, \alpha],$$

$$Tan(t0, b):[\text{fu}, \alpha], \quad Tan(t1, 0):[\text{em}, \alpha], \tag{7.72}$$

$$Tan(t2, 0):[\text{em}, \alpha], \quad Tan(t3, na):[\text{fu}, \alpha]. \tag{7.73}$$

Then all the sub-processes and valves are permitted to be locked or controlled. However the tank conditions (7.72) and (7.73) do not permit for setting processes $Pr_{2,3}$. The beer process Pr_0 can be verified to be set as follows:

- we have neither the forbiddance from locking sub-processes $SPr_{0,1}$, nor the forbiddance from controlling valve V_0 separate with beer in the normal direction, by EVALPSN clauses (7.34) and (7.37)–(7.39) and the input fact EVALP clauses;
- then we have the permission for locking sub-processes $SPr_{0,1}$, and controlling valve V_0 separate with beer in the normal direction and any liquid in the cross direction,

$$SPr(t0, v0, b) : [\text{f}, \gamma], \quad Val(v0, b, l) : [\text{cm}, \gamma],$$
$$SPr(v0, t1, b) : [\text{f}, \gamma],$$

where $l \in \{b, cw, ww, hw, na, cs, 0\}$, by EVALPSN clauses (7.35) and (7.42);

• moreover, we obtain the following EVALP clauses to represent tank conditions,

$$Tan(t0, b) : [\text{fu}, \alpha], \quad Tan(t1, 0) : [\text{em}, \alpha];$$

• thus we have the permission for setting the beer process Pr_0,

$$Pr(p0, b) : [\text{xs}, \gamma],$$

by EVALPSN clause (7.44).

According to the process schedule, the beer process Pr_0 has to start first, then the nitric acid process Pr_1 has to be verified its safety and processed parallel to process Pr_0 as soon as possible. We show the safety verification for process Pr_1 at the next stage.

2nd Stage The beer process Pr_0 has already started but not finished yet, then in order to verify the safety for processes $Pr_{1,2,3}$, the following fact EVALP clauses are input to the pipeline safety control EVALPSN:

$$SPr(t0, v0, b) : [1, \alpha], \quad Val(v0, b, 0) : [\text{cs}, \alpha],$$
$$SPr(v0, t1, b) : [1, \alpha], \quad Val(v1, 0, 0) : [\text{cs}, \alpha],$$
$$SPr(v0, t2, 0) : [\text{f}, \alpha],$$
$$SPr(v1, v0, 0) : [\text{f}, \alpha],$$
$$SPr(t3, v1, 0) : [\text{f}, \alpha],$$
$$Tan(t2, 0) : [\text{em}, \alpha], \quad Tan(t3, na) : [\text{fu}, \alpha].$$

The above tank conditions permit neither the cold water process Pr_2 nor the beer process Pr_3 to be set. We show that only the nitric acid process Pr_1 can be verified to be set as follows:

• we have neither the forbiddance from locking three sub-processes $SPr_{2,3,4}$, the forbiddance from controlling valves V_0 separate with any liquid in the normal direction and nitric acid in the cross direction, nor the forbiddance from controlling valve V_1 mix(open) with nitric acid in the normal direction and no liquid in the cross direction, by EVALPSN clauses (7.34) and (7.36)–(7.41) and the above fact EVALP clauses;

• therefore we have the permission for locking sub-processes $SPr_{2,3,4}$, and controlling valves V_0 and V_1 as described before,

$$SPr(v0, t2, na):[\mathtt{f}, \gamma], \quad Val(v0, b, na):[\mathtt{cm}, \gamma],$$
$$SPr(v1, v0, na):[\mathtt{f}, \gamma], \quad Val(v1, na, 0):[\mathtt{cs}, \gamma],$$
$$SPr(t3, v1, na):[\mathtt{f}, \gamma],$$

by EVALPSN clauses (7.35), (7.42) and (7.43);

• moreover we have the tank conditions,

$$Tan(t3, na):[\mathtt{fu}, \alpha], \quad Tan(t2, 0):[\mathtt{em}, \alpha];$$

• thus we have the permission for setting the nitric acid process Pr_1,

$$Pr(p1, na):[\mathtt{xs}, \gamma],$$

by EVALPSN clause (7.45).

Both the beer process Pr_0 and the nitric acid process Pr_1 have already started, then processes $Pr_{2,3}$ have to be verified their safety. We will show it at the next stage.

3rd Stage In order to verify the safety for the cold water process Pr_2 and the beer process Pr_3, the following fact EVALP clauses are input to the pipeline safety control EVALPSN:

$$SPr(t0, v0, b):[\mathtt{1}, \alpha], \quad Val(v0, b, na):[\mathtt{cs}, \alpha],$$
$$SPr(v0, t1, b):[\mathtt{1}, \alpha], \quad Val(v1, na, 0):[\mathtt{cm}, \alpha],$$
$$SPr(v0, t2, na):[\mathtt{1}, \alpha],$$
$$SPr(v1, v0, na):[\mathtt{1}, \alpha],$$
$$SPr(t3, v1, na):[\mathtt{1}, \alpha].$$

Apparently, neither the cold water process Pr_2 nor the beer process Pr_3 is permitted to be set, since there is no tank condition in the input fact EVALP clauses. We show the safety verification for process Pr_2 as an example:

• we have the forbiddance from locking sub-processes $SPr_{2,3,4}$, the forbiddance from controlling valve V_0 separate with beer in the normal direction and cold water in the cross direction, and the forbiddance from controlling valve V_1 mix with cold water in the normal direction and no liquid in the cross direction,

$$SPr(v0, t2, cw):[\mathtt{f}, \beta], \quad Val(v0, b, cw):[\mathtt{cm}, \beta],$$
$$SPr(v1, v0, cw):[\mathtt{f}, \beta], \quad Val(v1, cw, 0):[\mathtt{cs}, \beta],$$
$$SPr(t3, v1, cw):[\mathtt{f}, \beta],$$

by EVALPSN clauses (7.34), (7.39) and (7.40) and the input fact EVALP clauses.

The finish condition for the nitric acid process Pr_1 is that tank T_2 is fully filled with nitric acid and tank T_3 is empty. If the nitric acid process Pr_1 has finished and its finish conditions are satisfied, process Pr_1 is permitted to be released (unset) and has been released in fact. It is also supposed that tank T_3 is filled with cold water and tank T_2 is empty as preparation for the cold water process Pr_2 immediately after process Pr_1 has finished. Then the cold water process Pr_2 has to be verified and start according to the process schedule. We show the safety verification for the cold water process Pr_2 at the next stage.

4th Stage If the nitric acid process Pr_1 has finished and its finishing condition is satisfied, sub-processes $SPr_{2,3,4}$, valve V_1, and the cross direction of valve V_0 are permitted to be released by EVALP clauses (7.52)–(7.57). They have been released in fact. Then, since only the beer process Pr_0 is being processed, other three processes $Pr_{1,2,3}$ have to be verified. In order to do that, the following fact EVALP clauses are input to the pipeline safety control EVALPSN:

$$SPr(t0, v0, b):[1, \alpha], \quad Val(v0, b, 0):[\text{cs}, \alpha],$$
$$SPr(v0, t1, b):[1, \alpha], \quad Val(v1, 0, 0):[\text{cs}, \alpha],$$
$$SPr(v0, t2, 0):[\text{f}, \alpha],$$
$$SPr(v1, v0, 0):[\text{f}, \alpha],$$
$$SPr(t3, v1, 0):[\text{f}, \alpha],$$
$$Tan(t2, 0):[\text{em}, \alpha], \quad Tan(t3, cw):[\text{fu}, \alpha].$$

Since the beer process Pr_0 is still being processed, neither the nitric acid process Pr_1 nor the beer process Pr_3 is permitted to be set, and only the cold water process Pr_2 is permitted to be set as well as process Pr_1 at the 2nd stage. Therefore we have the permission for setting the cold water process Pr_2,

$$Pr(p2, cw):[\text{xs}, \gamma]$$

by EVALPSN clause (7.46).

Now, both the beer process Pr_0 and the cold water process Pr_2 are being processed. Then apparently any other processes are not permitted to be set. Moreover even if one of processes $Pr_{0,2}$ has finished, the beer process Pr_3 is not permitted to be set until both processes $Pr_{0,2}$ have finished. We show the safety verification for the beer process Pr_3 at the following stages.

5th Stage If neither the beer process Pr_0 nor the cold water process Pr_2 has finished, we have to verify the safety for the nitric acid process Pr_1 and the beer process Pr_3. The following fact EVALP clauses are input to the pipeline safety control EVALPSN:

$$SPr(t0, v0, b):[1, \alpha], \quad Val(v0, b, cw):[\text{cs}, \alpha],$$
$$SPr(v0, t1, b):[1, \alpha], \quad Val(v1, cw, 0):[\text{cm}, \alpha],$$
$$SPr(v0, t2, cw):[1, \alpha],$$
$$SPr(v1, v0, cw):[1, \alpha],$$
$$SPr(t3, v1, cw):[1, \alpha].$$

Then, since all the sub-processes and valves are locked and controlled, neither processes Pr_1 nor Pr_3 is permitted to be set. It is shown that the beer process Pr_3 is not permitted to be set as follows:

- we have the forbiddance from locking sub-processes $SPr_{2,3,4}$ in process Pr_3 and controlling valves $V_{0,1}$,

$$SPr(v0, t2, b) : [\mathtt{f}, \beta], \quad Val(v0, b, b) : [\mathtt{cs}, \beta],$$
$$SPr(t3, v1, b) : [\mathtt{f}, \beta], \quad Val(v1, b, 0) : [\mathtt{cs}, \beta],$$
$$SPr(v1, v0, b) : [\mathtt{f}, \beta],$$

by EVALPSN clauses (7.34), (7.36) and (7.40) and the input fact EVALP clauses;
- therefore we cannot derive the permission for setting process Pr_3.

The finish condition for the cold water process Pr_2 is that tank T_2 is fully filled with cold water and tank T_3 is empty. If the cold water process Pr_2 has finished and its finish condition is satisfied, process Pr_2 is permitted to be released(unset). It is also supposed that tank T_3 is filled with beer and tank T_2 is empty as preparation for the beer process Pr_3 immediately after process Pr_2 has finished. Then the beer process Pr_3 has to be verified and start according to the process schedule, but the beer process Pr_3 cannot be permitted to be set. We show the safety verification for process Pr_3 at the next stage.

6th Stage If the cold water process Pr_2 has finished and its finish condition is satisfied, the three sub-processes $SPr_{2,3,4}$, the valve V_1, and the cross direction of the valve V_0 are permitted to be released by EVALP clauses (7.52)–(7.57). They have been released in fact. Then, since only the beer process Pr_0 is being processed, processes $Pr_{1,2,3}$ have to be verified their safety. In order to do that, the following fact EVALP clauses are input to the pipeline safety control EVALPSN:

$$SPr(t0, v0, b) : [1, \alpha], \quad Val(v0, b, 0) : [\mathtt{cs}, \alpha],$$
$$SPr(v0, t1, b) : [1, \alpha], \quad Val(v1, 0, 0) : [\mathtt{cs}, \alpha],$$
$$SPr(v0, t2, 0) : [\mathtt{f}, \alpha],$$
$$SPr(v1, v0, 0) : [\mathtt{f}, \alpha],$$
$$SPr(t3, v1, 0) : [\mathtt{f}, \alpha],$$
$$Tan(t2, 0) : [\mathtt{em}, \alpha], \quad Tan(t3, b) : [\mathtt{fu}, \alpha].$$

Since the beer process Pr_0 is still being processed, the beer process Pr_3 is not verified its safety due to the tank conditions and safety property **Val** for valve V_0. The safety verification is carried out as follows:

- we have the forbiddance from controlling valve V_0 mix,

$$Val(v0, b, b) : [\mathtt{cs}, \beta],$$

by EVALPSN clause (7.36);

- therefore we cannot have the permission for setting the beer process Pr_3 then. On the other hand, even if the beer process Pr_0 has finished with the cold water process Pr_2 still being processed, the beer process Pr_3 is not permitted to be set. If both processes $Pr_{0,2}$ have finished, the beer process Pr_3 is assured its safety and set. Then process Pr_3 starts according to the process schedule. We omit the rest of the safety verification stages.

7.5 Before-After EVALPSN

In this section, we review an extended version of EVALPSN named bf(before-after)-EVALPSN formally, which can deal with before-after relations between two processes(time intervals) and introduce how to implement bf-EVALPSN aiming at the real-time safety verification for process order control [40, 41].

7.5.1 Before-After Relation in EVALPSN

First of all, we introduce a special literal $R(pi, pj, t)$ whose vector annotation represents the before-after relation between processes $Pr_i(pi)$ and $Pr_j(pj)$ at time t, where processes can be regarded as time intervals in general, and literal $R(pi, pj, t)$ is called a *bf-literal*.[3]

Definition 5.1 (*bf-EVALPSN*) An extended vector annotated literal $R(p_i, p_j, t)$: $[\mu_1, \mu_2]$ is called a *bf-EVALP* literal, where μ_1 is a vector annotation and $\mu_2 \in \{\alpha, \beta, \gamma\}$. If an EVALPSN clause contains bf-EVALP literals, it is called a *bf-EVALPSN clause* or just a *bf-EVALP clause* if it contains no strong negation. A bf-EVALPSN is a finite set of EVALPSN clauses and bf-EVALPSN clauses.

We provide some paraconsistent interpretations of vector annotations for representing bf-relations, which are called *bf-annotations*. Strictly speaking, bf-relations between time intervals are classified into 15 sorts according to bf-relations between start/finish times of two time intervals. We define the 15 sorts of bf-relations in bf-EVALPSN with regarding processes as time intervals.

Suppose that there are two processes, Pr_i with its start/finish times x_s and x_f, and Pr_j with its start/finish times y_s and y_f.

Before (be)/**After** (af)

Firstly, we define the most basic bf-relations *before/after* according to the bf-relation between each start time of two processes, which are represented by bf-annotations be/af, respectively. If one process has started before/after another one starts, then the bf-relations between those processes are defined as "before(be)/after(af)", respectively. The bf-relations also are described in Fig. 7.10 with the condition that process

---

[3]Hereafter, expression "**before-after**" is abbreviated as just "bf" in this chapter.

Fig. 7.10 Bf-relations, before/after

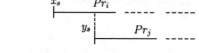

Fig. 7.11 Bf-relations, disjoint before/after

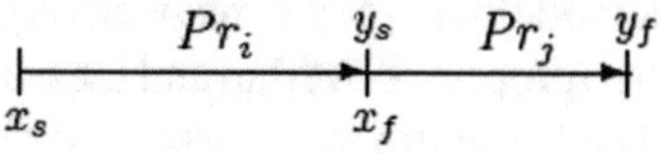

Fig. 7.12 Bf-relations, immediate before/after

Pr_i has started before process Pr_j starts. The bf-relation between their start/finish times is denoted by the inequality $\{x_s < y_s\}$.[4] For example, a fact at time t "process Pr_i has started before process Pr_j starts" can be represented by bf-EVALP clause

$$R(pi, pj, t) : [\text{be}, \alpha].$$

Disjoint Before (db)/After (da)

Bf-relations *disjoint before/after* between processes Pr_i and Pr_j are represented by bf-annotations db/da, respectively. The expression "disjoint before/after" implies that there is a timelag between the earlier process finish time and the later one start time. They are also described in Fig. 7.11 with the condition that process Pr_i has finished before process Pr_j starts. The bf-relation between their start/finish times is denoted by the inequality $\{x_f < y_s\}$. For example, an obligation at time t "process Pr_i must start after process Pr_j has finished" can be represented by bf-EVALP clause

$$R(pi, pj, t) : [\text{da}, \beta].$$

Immediate Before (mb)/After (ma)

Bf-relations *immediate before/after* between processes Pr_i and Pr_j are represented by bf-annotations mb/ma, respectively. The expression "immediate before/after" implies that there is no timelag between the earlier process finish time and the later one start time. The bf-relations are also described in Fig. 7.12 with the condition that process Pr_i has finished immediately before process Pr_j starts. The bf-relation between their start/finish times is denoted by the equality $\{x_f = y_s\}$. For example, a fact at time t "process Pr_i has finished immediately before process Pr_j starts" can be represented by bf-EVALP clause

$$R(pi, pj, t) : [\text{mb}, \alpha].$$

Joint Before (jb)/After (ja)

---

[4]If time t_1 is earlier than time t_2, we conveniently denote the before-after relation by the inequality $t_1 < t_2$.

Fig. 7.13 Bf-relations, joint before/after

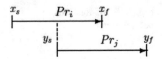

Bf-relations *joint before/after* between processes Pr_i and Pr_j are represented by bf-annotations jb/ja, respectively. The expression "joint before/after" imply that the two processes overlap and the earlier process has finished before the later one finishes. The bf-relations are also described in Fig. 7.13 with the condition that process Pr_i has started before process Pr_j starts and process Pr_i has finished before process Pr_j finishes. The bf-relation between their start/finish times is denoted by the inequalities $\{x_s < y_s < x_f < y_f\}$. For example, a fact at time t "process Pr_i has started before process Pr_j starts and finished before process Pr_j finishes" can be represented by bf-EVALP clause

$$R(pi, pj, t):[\text{jb}, \alpha].$$

S-included Before(sb)/After(sa)

Bf-relations *s-included before/after* between processes Pr_i and Pr_j are represented by bf-annotations sb/sa, respectively. The expression "s-included before/after" implies that one process has started before another one starts and they finish at the same time. The bf-relations are also described in Fig. 7.14 with the condition that process Pr_i has started before process Pr_j starts and they finish at the same time. The bf-relation between their start/finish times is denoted by the equality and inequalities $\{x_s < y_s < x_f = y_f\}$. For example, a fact at time t "process Pr_i has started before process Pr_j starts and they finish at the same time" can be represented by bf-EVALP clause

$$R(pi, pj, t):[\text{sb}, \alpha].$$

Included Before (ib)/After (ia)

Bf-relations *included before/after* between processes Pr_i and Pr_j are represented by bf-annotations ib/ia, respectively. The expression "included before/after" implies that one process has started before another one starts and the earlier one finishes after another one has finished. The bf-relations are also described in Fig. 7.15 with the condition that process Pr_i has started before process Pr_j starts and finishes after process Pr_j has finished. The bf-relation between their start/finish times is denoted by the inequalities $\{x_s < y_s, y_f < x_f\}$.

Fig. 7.14 Bf-relations, S-included before/after

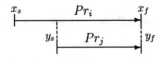

Fig. 7.15 Bf-relations, included before/after

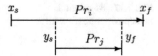

For example, an obligation at the time t "process Pr_i must start before process Pr_j starts and process Pr_i must finish after process Pr_j has finished" can be represented by bf-EVALP clause

$$R(pi, pj, t):[\mathtt{ib}, \beta].$$

F-included Before (fb)/After (fa)

The bf-relations *f-include before/after* between processes Pr_i and Pr_j are represented by bf-annotations fb/fa, respectively. The expression "f-included before/after" implies that two processes have started at the same time and one process has finished before another one finishes. The bf-relations are also described in Fig. 7.16 with the condition that processes Pr_i and Pr_j have started at the same time and process Pr_i finishes after process Pr_j has finished. The bf-relation between their start/finish times is denoted by the equality and inequality $\{x_s = y_s, \ y_f < x_f\}$. For example, a fact at time t "processes Pr_i and Pr_j have started at the same time and process Pr_i has finished after process Pr_j finished" can be represented by bf-EVALP clause

$$R(pi, pj, t):[\mathtt{fa}, \alpha].$$

Paraconsistent Before-after (pba)

Bf-relation *paraconsistent before-after* between processes Pr_i and Pr_j is represented by bf-annotation pba. The expression "paraconsistent before-after" implies that the two processes have started at the same time and also finished at the same time. The bf-relation is also described in Fig. 7.17 with the condition that processes Pr_i and Pr_j have not only started but also finished at the same time. The bf-relation between their start/finish times is denoted by the equalities $\{x_s = y_s, \ y_f = x_f\}$.

For example, an obligation at time t "processes Pr_i and Pr_j must not only start but also finish at the same time" can be represented by bf-EVALP clause

$$R(pi, pj, t):[\mathtt{pba}, \beta].$$

Here we define the epistemic negation $\neg_1$ that maps bf-annotations to themselves in bf-EVALPSN.

Fig. 7.16 Bf-relations, F-included before/after

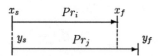

Fig. 7.17 Bf-relation,
paraconsistent before-after

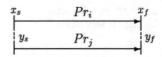

Definition 5.2 (*Epistemic Negation* $\neg_1$ *for Bf-annotations*)
The epistemic negation $\neg_1$ over the set of bf-annotations,

$$\{be, af, da, db, ma, mb, ja, jb, sa, sb, ia, ib, fa, fb, pba\}$$

is obviously defined as the following mapping:

$$\neg_1(af) = be, \quad \neg_1(be) = af,$$
$$\neg_1(da) = db, \quad \neg_1(db) = da,$$
$$\neg_1(ma) = mb, \quad \neg_1(mb) = ma,$$
$$\neg_1(ja) = jb, \quad \neg_1(jb) = ja,$$
$$\neg_1(sa) = sb, \quad \neg_1(sb) = sa,$$
$$\neg_1(ia) = ib, \quad \neg_1(ib) = ia,$$
$$\neg_1(fa) = fb, \quad \neg_1(fb) = fa,$$
$$\neg_1(pba) = pba.$$

If we consider before-after measure over the meaningful 15 bf-annotations, obviously there exists a partial order($<_h$) based on the before-after measure, where $\mu_1 <_h \mu_2$ is intuitively interpreted that bf-annotation μ_1 denotes a more "before" degree than bf-annotation μ_2, and $\mu_1, \mu_2 \in \{be, af, db, da, mb, ma, jb, ja, ib, ia, sb, sa, fb, fa, pba\}$. If $\mu_1 <_h \mu_2$ and $\mu_2 <_h \mu_1$, we denote it $\mu_1 \equiv_h \mu_2$. Then we obtain the following ordering:

$$db <_h mb <_h jb <_h sb <_h ib <_h fb <_h pba <_h ia <_h ja <_h ma <_h da$$
$$\text{and}$$
$$sb \equiv_h be <_h af \equiv_h sa.$$

On the other hand, if we take before-after knowledge (information) amount of each bf-relation into account as another measure, obviously there also exists another partial order($<_v$) in terms of before-after knowledge amount, where $\mu_1 <_v \mu_2$ is intuitively interpreted that bf-annotation μ_1 has less knowledge amount in terms of bf-relation than bf-annotation μ_2. If $\mu_1 <_v \mu_2$ and $\mu_2 <_v \mu_1$, we denote it $\mu_1 \equiv_v \mu_2$. Then we obtain the following ordering:

$$\text{be} <_v \mu_1, \quad \mu_1 \in \{ \text{db}, \text{mb}, \text{jb}, \text{sb}, \text{ib} \},$$
$$\text{af} <_v \mu_2, \quad \mu_1 \in \{ \text{da}, \text{ma}, \text{ja}, \text{sa}, \text{ia} \},$$
$$\text{db} \equiv_v \text{mb} \equiv_v \text{jb} \equiv_v \text{sb} \equiv_v \text{ib} \equiv_v \text{fb} \equiv_v \text{pba} \equiv_v$$
$$\text{fa} \equiv_v \text{ia} \equiv_v \text{sa} \equiv_v \text{ja} \equiv_v \text{ma} \equiv_v \text{da}$$

and

$$\text{be} \equiv_v \text{af}.$$

If we take the before-after degree as the horizontal measure and the before-after knowledge amount as the vertical one, we obtain the complete bi-lattice $\mathcal{T}_v(12)_{bf}$ of vector annotations that includes the 15 bf-annotations.

$$\begin{aligned}
\mathcal{T}_v(12)_{bf} = \{ & \bot_{12}(0, 0), \ldots, \text{be}(0, 8), \ldots, \text{db}(0, 12), \ldots, \text{mb}(1, 11), \ldots, \\
& \text{jb}(2, 10), \ldots, \text{sb}(3, 9), \ldots, \text{ib}(4, 8), \ldots, \text{fb}(5, 7), \ldots, \\
& \text{pba}(6, 6), \ldots, \text{fa}(7, 5), \ldots, \text{af}(8, 0), \ldots, \text{ia}(8, 4), \ldots, \\
& \text{sa}(9, 3), \ldots, \text{ja}(10, 2), \ldots, \text{ma}(11, 1), \ldots, \text{da}(12, 0), \ldots, \\
& \top_{12}(12, 12) \},
\end{aligned}$$

which is described as the Hasse's diagram in Fig. 7.18.

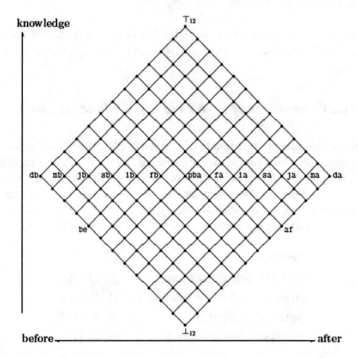

Fig. 7.18 The complete bi-lattice $\mathcal{T}_v(12)_{bf}$ of bf-annotations

We note that bf-EVALP literal

$R(pi, pj, t) : [\mu_1(m, n), \mu_2]$,

where $\mu_2 \in \{\alpha, \beta, \gamma\}$ and

$\mu_1 \in \{\text{be}, \text{db}, \text{mb}, \text{jb}, \text{sb}, \text{ib}, \text{fb}, \text{pba}, \text{fa}, \text{ia}, \text{sa}, \text{jb}, \text{ma}, \text{da}, \text{af}\}$,

is not well annotated if $m \neq 0$ and $n \neq 0$, however, the bf-EVALP literal is equivalent to the following two well annotated bf-EVALP literals:

$$R(pi, pj) : [(m, 0), \mu] \quad \text{and} \quad R(pi, pj) : [(0, n), \mu].$$

Therefore such a non-well annotated bf-EVALP literal can be regarded as the conjunction of two well annotated EVALP literals.

For example, suppose that there is a non-well annotated bf-EVALP clause,

$$R(pi, pj, t_1) : [(k, l), \mu_1] \rightarrow R(pi, pj, t_2) : [(m, n), \mu_2],$$

where $k \neq 0$, $l \neq 0$, $m \neq 0$ and $n \neq 0$. It can be equivalently transformed into the following two well annotated bf-EVALP clauses,

$$R(pi, pj, t_1) : [(k, 0), \mu_1] \wedge R(pi, pj, t_1) : [(0, l), \mu_1] \rightarrow R(pi, pj, t_2) : [(m, 0), \mu_2],$$
$$R(pi, pj, t_1) : [(k, 0), \mu_1] \wedge R(pi, pj, t_2) : [(0, l), \mu_1] \rightarrow R(pi, pj, t_2) : [(0, n), \mu_2].$$

7.5.2 Implementation of Bf-EVALPSN Verification System

In this subsection we introduce how to implement bf-EVALPSN based process order safety verification with a simple example. For simplicity, we do not consider cases in which one process starts/finishes with another one starts/finishes at the same time, then the process order control system can deal with before-after relations more simply, which means that bf-annotations(relations) sb/sa, fb/fa and pba are excluded.

We take the following ten bf-annotations with vector annotations into account:

before(**be**)/after(af),	$(0, 4)/(4, 0)$,
discrete before(**db**)/after(da),	$(0, 7)/(7, 0)$,
immediate before(**mb**)/after(ma),	$(1, 6)/(6, 1)$,
joint before(**jb**)/after(ja),	$(2, 5)/(5, 2)$,
included before(**ib**)/after(ia).	$(3, 4)/(4, 3)$.

The complete bi-lattice $\mathcal{T}_v(7)_{bf}$ including the ten bf-annotations is described as the Hasse's diagram in Fig. 7.19.

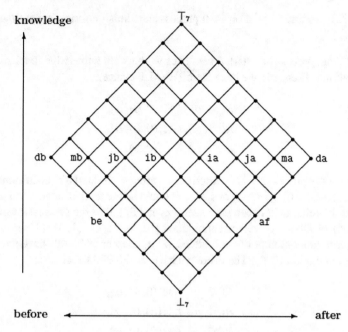

Fig. 7.19 The complete bi-lattice $\mathcal{T}_v(7)_{bf}$ of bf-annotations

Now we show an example of implementing the real-time process order safety verification system in bf-EVALPSN.

Example 5.1 We suppose that there are four processes Pr_0(id $p0$), Pr_1(id $p1$) Pr_2(id $p2$) and the next process Pr_3(id $p3$) not appearing in Fig. 7.20. Those processes are supposed to be processed according to the processing schedule in Fig. 7.20.

Then we consider three bf-relations represented by the following bf-EVALP clauses (7.74)–(7.76):

$$R(p0, p1, t_i) : [(i_1, j_1), \alpha], \tag{7.74}$$

$$R(p1, p2, t_i) : [(i_2, j_2), \alpha], \tag{7.75}$$

$$R(p2, p3, t_i) : [(i_3, j_3), \alpha], \tag{7.76}$$

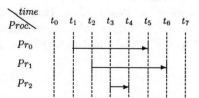

Fig. 7.20 Process timing chart

which will be inferred based on each process start/finish information at time t_i ($i =$ 0, 1, 2, . . . , 7).

At time t_0, no process has started yet. Thus we have no knowledge in terms of any bf-relations. Therefore we have the bf-EVALP clauses,

$$R(p0, p1, t_0) : [(0, 0), \alpha],$$
$$R(p1, p2, t_0) : [(0, 0), \alpha],$$
$$R(p2, p3, t_0) : [(0, 0), \alpha].$$

At time t_1, only process Pr_0 has started before process Pr_1 starts, Then bf-annotations db(0, 7), mb(1, 6), jb(2, 5) or ib(3, 4) could be the final bf-annotation to represent the bf-relation between processes Pr_0 and Pr_1, thus the greatest lower bound be(0, 4) of the set of vector annotations {(0, 7), (1, 6), (2, 5), (3, 4)} becomes the vector annotation of bf-literal $R(p0, p1, t_1)$. Other bf-literals have the bottom vector annotation (0, 0). Therefore we have the bf-EVALP clauses,

$$R(p0, p1, t_1) : [(0, 4), \alpha],$$
$$R(p1, p2, t_1) : [(0, 0), \alpha],$$
$$R(p2, p3, t_1) : [(0, 0), \alpha].$$

At time t_2, the second process Pr_1 also has started before process Pr_0 finishes. Then bf-annotations jb (2, 5) or ib (3, 4) could be the final bf-relation to represent the bf-relation between processes Pr_0 and Pr_1. Thus the greatest lower bound (2, 4) of the set of vector annotations {(2, 5), (3, 4)} has to be the vector annotation of bf-literal $R(p0, p1, t_2)$. In addition, bf-literal $R(p1, p2, t_2)$ has bf-annotation be(0, 4) as well as bf-literal $R(p0, p1, t_1)$ since process Pr_1 has also started before process Pr_2 starts. On the other hand, bf-literal $R(p2, p3, t2)$ has the bottom vector annotation (0, 0) since process Pr_3 has not started yet. Therefore we have the bf-EVALP clauses,

$$R(p0, p1, t_2) : [(2, 4), \alpha],$$
$$R(p1, p2, t_2) : [(0, 4), \alpha],$$
$$R(p2, p3, t_2) : [(0, 0), \alpha].$$

At time t_3, process Pr_2 has started before both processes Pr_0 and Pr_1 finish. Then both bf-literals $R(p0, p1, t_3)$ and $R(p1, p2, t_3)$ have the same vector annotation (2, 4) as well as bf-literal $R(p0, p1, t_2)$. Moreover bf-literal $R(p2, p3, t_3)$ has bf-annotation be(0, 4) as well as bf-literal $R(p0, p1, t_1)$. Therefore we have the bf-EVALP clauses,

$$R(p0, p1, t_3) : [(2, 4), \alpha],$$
$$R(p1, p2, t_3) : [(2, 4), \alpha],$$
$$R(p2, p3, t_3) : [(0, 4), \alpha].$$

At time t_4, process Pr_2 has finished before both processes Pr_0 and Pr_1 finish. Then bf-literal $R(p0, p1, t_4)$ still has the same vector annotation $(2, 4)$ as well as the previous time t_3. In addition bf-literal $R(p1, p2, t_4)$ has its final bf-annotation $\text{ib}(3, 4)$. For the final bf-relation between processes Pr_2 and Pr_3 there are still two alternatives: (1) if process Pr_3 starts immediately after process Pr_2 has finished, bf-literal $R(p2, p3, t_4)$ has its final bf-annotation $\text{mb}(1, 6)$; (2) if process Pr_3 does not start immediately after process Pr_2 has finished, bf-literal $R(p2, p3, t_4)$ has its final bf-annotation $\text{db}(0, 7)$. Either way, we have the knowledge that process Pr_2 has just finished at time t_4, which can be represented by vector annotation $(0, 6)$ that is the greatest lower bound of the set of vector annotations $\{(1, 6), (0, 7)\}$. Therefore we have the bf-EVALP clauses,

$$R(p1, p2, t_4) : [(2, 4), \alpha],$$
$$R(p2, p3, t_4) : [(3, 4), \alpha],$$
$$R(p3, p4, t_4) : [(0, 6), \alpha].$$

At time t_5, process Pr_0 has finished before processes Pr_1 finishes. Then bf-literal $R(p0, p1, t_5)$ has its final bf-annotation $\text{jb}(2, 5)$, and bf-literal $R(p2, p3, t_5)$ also has its final bf-annotation $\text{jb}(0, 7)$ because process Pr_3 has not started yet. Therefore we have the bf-EVALP clauses,

$$R(p1, p2, t_5) : [\text{jb}(2, 5), \alpha],$$
$$R(p2, p3, t_5) : [\text{ib}(3, 4), \alpha],$$
$$R(p3, p4, t_5) : [\text{db}(0, 7), \alpha],$$

and all the bf-relations have been determined at time t_5 before process Pr_1 finishes and process Pr_3 starts.

In Example 5.1, we have shown how the vector annotations of bf-literals are updated according to the start/finish information of processes in real-time. We will introduce the real-time safety verification for process order control based on bf-EVALPSN with small examples in the subsequent subsection.

7.5.3   Safety Verification in Bf-EVALPSN

We present the basic idea of bf-EVALPSN safety verification for process order with a simple example.

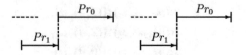

Fig. 7.21 Process schedule example

Suppose that two processes Pr_0 and Pr_1 are being processed repeatedly, and process Pr_1 must be processed immediately before process Pr_0 starts as shown in Fig. 7.21.

In bf-EVALPSN process order safety verification, the safety for process order is verified based on the safety properties to be assured in the process schedule. In order to verify the safety for the process order in Fig. 7.21, we assume two safety properties **SP-0** and **SP-1** for processes Pr_0 and Pr_1, respectively:

SP-0 process Pr_0 must start immediately after process Pr_1 has finished,
SP-1 process Pr_1 must start in a while after (disjoint after) process Pr_0 has finished.

Then safety properties **SP-0** and **SP-1** should be verified immediately before processes Pr_0 and Pr_1 start, respectively.

In order to verify the bf-relation "immediate after" with safety property **SP-0**, it should be verified whether process Pr_1 has finished immediately before process Pr_0 starts or not, and the safety verification should be carried out immediately after process Pr_1 has finished. Then bf-literal $R(p0, p1, t)$ must have vector annotation $(6, 0)$, which means that process Pr_1 has finished but process Pr_0 has not started yet. Therefore safety property **SP-0** is translated to the bf-EVALPSN-clauses,

SP-0

$$R(p0, p1, t) : [(6, 0), \alpha] \wedge \sim R(p0, p1, t) : [(7, 0), \alpha]$$
$$\rightarrow st(p0, t) : [\mathtt{f}(0, 1), \gamma], \tag{7.77}$$
$$\sim st(p0, t) : [\mathtt{f}(0, 1), \gamma] \rightarrow st(p0, t) : [\mathtt{f}(0, 1), \beta], \tag{7.78}$$

where literal $st(pi, t)$ represents "process Pr_i starts at time t" and the set of its vector annotations constitutes the complete lattice

$$\mathcal{T}_v(1) = \{\perp(0, 0), \mathtt{t}(1, 0), \mathtt{f}(1, 0), \top(1, 1)\}.$$

For example, EVALP-clause $st(p0, t) : [\mathtt{f}(0, 1), \gamma]$ can be informally interpreted as "it is permitted for process Pr_0 to start at time t".

On the other hand in order to verify bf-relation "disjoint after" with safety property **SP-1**, it should be verified whether there is a timelag between process Pr_0 finish time and process Pr_1 start time or not. Then bf-literal $R(p1, p0, t)$ must have bf-annotation $\mathtt{da}(7, 0)$. Therefore safety property **SP-1** is translated into the following bf-EVALPSN clauses:

SP-1

$$R(p1, p0, t) : [(7, 0), \alpha] \rightarrow Start(p1, t) : [(0, 1), \gamma], \tag{7.79}$$

$$\sim Start(p1, t) : [(0, 1), \gamma] \rightarrow Start(p1, t) : [(0, 1), \beta]. \tag{7.80}$$

We show how to verify the process order safety based on safety properties **SP-0** and **SP-1** in bf-EVALPSN. In order to verify the process order safety, the following safety verification cycle consisting of two steps is applied repeatedly.

Safety Verification Cycle

1st Step (safety verification for starting process Pr_1)
Suppose that process Pr_1 has not started yet at time t_1. If process Pr_0 has already finished at time t_1, we have the bf-EVALP clause,

$$R(p1, p0, t_1) : [(7, 0), \alpha]. \tag{7.81}$$

On the other hand, if process Pr_0 has just finished at time t_1, we have the bf-EVALP clause,

$$R(p1, p0, t_1) : [(6, 0), \alpha]. \tag{7.82}$$

If bf-EVALP clause (7.81) is input to safety property **SP-1** consisting of Bf-EVALPSN clauses (7.79) and (7.80), we obtain the EVALP clause,

$$st(p1, t_1) : [(0, 1), \gamma]$$

and the safety for starting process Pr_1 is assured. On the other hand, if bf-EVALP clause (7.82) is input to the same safety property **SP-1**, we obtain the EVALP clause

$$st(p1, t_1) : [(0, 1), \beta],$$

then the safety for starting process Pr_1 is not assured.
2nd Step (safety verification for starting process Pr_0)
Suppose that process Pr_0 has not started yet at time t_2. If process Pr_1 has just finished at time t_2, we have the bf-EVALP clause,

$$R(p0, p1, t_2) : [(6, 0), \alpha]. \tag{7.83}$$

On the other hand, if process Pr_1 has not finished yet at time t_2, we have the bf-EVALP clause,

$$R(p0, p1, t_2) : [(4, 0), \alpha]. \tag{7.84}$$

If bf-EVALP clause (7.83) is input to safety property **SP-0** {(7.77) and (7.78)}, we obtain the EVALP clause,

$$st(p0, t_2):[(0, 1), \gamma],$$

and the safety for starting process Pr_0 is assured. On the other hand, if bf-EVALP clause (7.84) is input to the same safety property **SP-0**, we obtain the EVALP clause,

$$st(p1, t):[(0, 1), \beta],$$

then the safety for starting process Pr_0 is not assured.

Example 5.2 In this example we suppose the same pipeline network as shown in Fig. 7.4 and the same process schedule as shown in Fig. 7.7.

process Pr_0, a brewery process using
    line-1, tank $T_0 \longrightarrow$ valve $V_0 \longrightarrow$ tank T_1;
process Pr_1, a cleaning process by nitric acid using
    line-2, tank $T_3 \longrightarrow$ valve $V_1 \longrightarrow$ Valve $V_0 \longrightarrow$ tank T_2;
process Pr_2, a cleaning process by water in line-1;
process Pr_3, a brewery process using both line-1 and line-2 with mixing at valve V_0

The above four processes are supposed to be processed according to the following processing schedule. We assume the following four process order safety properties for each process:

SP-2 process Pr_0 must start before any other processes start;
SP-3 process Pr_1 must start immediately after process Pr_0 has started;
SP-4 process Pr_2 must start immediately after process Pr_1 has finished;
SP-5 process Pr_3 must start immediately after both processes Pr_0 and Pr_2 have finished.

Safety property **SP-2** is translated into the bf-EVALPSN clauses,

> **SP-2**
> $\sim R(p0, p1, t):[(4, 0), \alpha] \wedge \sim R(p0, p2, t):[(4, 0), \alpha] \wedge$
> $\sim R(p0, p3, t):[(4, 0), \alpha] \rightarrow Start(p0, t):[(0, 1), \gamma],$
> $\sim Start(p0, t):[(0, 1), \gamma] \rightarrow Start(p0, t):[(0, 1), \beta].$ (7.85)

As well as safety property **SP-2**, other safety properties **SP-3**, **SP-4** and **SP-5** are also translated into the bf-EVALPSN clauses,

> **SP-3**
> $R(p1, p0, t):[(4, 0), \alpha] \rightarrow Start(p1, t):[(0, 1), \gamma],$
> $\sim Start(p1, t):[(0, 1), \gamma] \rightarrow Start(p1, t):[(0, 1), \beta],$ (7.86)

SP-4

$$R(p2, p1, t) : [(6, 0), \alpha] \wedge {\sim}R(p2, p1, t) : [(7, 0), \alpha]$$
$$\rightarrow Start(p2, t) : [(0, 1), \gamma],$$
$${\sim}Start(p2, t) : [(0, 1), \gamma] \rightarrow Start(p2, t) : [(0, 1), \beta], \qquad (7.87)$$

SP-5

$$R(p3, p0, t) : [(6, 0), \alpha] \wedge R(p3, p2, t) : [(6, 0), \alpha] \wedge$$
$${\sim}R(p3, p2, t) : [(7, 0), \alpha] \rightarrow Start(p3, t) : [(0, 1), \gamma],$$
$$R(p3, p0, t) : [(6, 0), \alpha] \wedge R(p3, p2, t) : [(6, 0), \alpha] \wedge$$
$${\sim}R(p3, p0, t) : [(7, 0), \alpha] \rightarrow Start(p3, t) : [(0, 1), \gamma],$$
$${\sim}Start(p3, t) : [(0, 1), \gamma] \rightarrow Start(p3, t) : [(0, 1), \beta]. \qquad (7.88)$$

We introduce the safety verification stages for the process order in Fig. 7.7 as follows.

Initial Stage (t_0) No process has started at time t_0, we have no information in terms of all bf-relations between all processes Pr_0, Pr_1, Pr_2 and Pr_3, thus we have the bf-EVALP clauses,

$$R(p0, p1, t_0) : [(0, 0), \alpha], \qquad (7.89)$$
$$R(p0, p2, t_0) : [(0, 0), \alpha], \qquad (7.90)$$
$$R(p0, p3, t_0) : [(0, 0), \alpha]. \qquad (7.91)$$

In order to verify the safety for starting the first process Pr_0, the bf-EVALP clauses (7.89)–(7.91) are input to safety property **SP-2** (7.85). Then we obtain the EVALP clause,

$$Start(p0, t_0) : [(0, 1), \gamma],$$

which expresses permission for starting process Pr_0, and its safety is assured at time t_0. Otherwise, it is not assured.

2nd Stage (t_1) Suppose that only process Pr_0 has already started at time t_1. Then we have the bf-EVALP clauses,

$$R(p1, p0, t_1) : [(4, 0), \alpha]. \qquad (7.92)$$

In order to verify the safety for starting the second process Pr_1, the bf-EVALP clause (7.92) is input to safety property **SP-3** (7.86). Then we obtain the EVALP clause,

$$Start(p1, t_1) : [(0, 1), \gamma],$$

and the safety for starting process Pr_1 is assured at time t_1. Otherwise, it is not assured.

3rd Stage (t_2) Suppose that processes Pr_0 and Pr_1 have already started, and neither of them has finished yet at time t_2. Then we have the bf-EVALP clauses,

$$R(p2, p0, t_2):[(4, 0), \alpha], \tag{7.93}$$

$$R(p2, p1, t_2):[(4, 0), \alpha]. \tag{7.94}$$

In order to verify the safety for starting the third process Pr_2, if EVALP clause (7.94) is input to safety property **SP-4** (7.87), then we obtain the EVALP clause,

$$Start(p2, t_2):[(0, 1), \beta],$$

and the safety for starting process Pr_2 is not assured at time t_2. On the other hand, if process Pr_1 has just finished at time t_2, then, we have the bf-EVALP clause,

$$R(p2, p1, t_2):[(6, 0), \alpha]. \tag{7.95}$$

If bf-EVALP clause (7.95) is input to safety property **SP-4** (7.87), then we obtain the EVALP clause,

$$Start(p2, t_2):[(0, 1), \gamma],$$

and the safety for starting process Pr_2 is assured.

4th Stage (t_3)   Suppose that processes Pr_0, Pr_1 and Pr_2 have already started, processes Pr_0 and Pr_1 have already finished, and only process Pr_3 has not started yet at time t_3. Then we have the bf-EVALP clauses,

$$R(p3, p0, t_3):[(7, 0), \alpha], \tag{7.96}$$

$$R(p3, p1, t_3):[(7, 0), \alpha], \tag{7.97}$$

$$R(p3, p2, t_3):[(4, 0), \alpha]. \tag{7.98}$$

In order to verify the safety for starting the last process Pr_3, if bf-EVALP clauses (7.96) and (7.98) are input to safety property **SP-5** (7.88), then we obtain the EVALP clause,

$$Start(p3, t_3):[(0, 1), \beta],$$

and the safety for starting process Pr_3 is not assured at time t_3. On the other hand, if process Pr_2 has just finished at time t_3, then we have the bf-EVALP clause,

$$R(p3, p2, t_3):[(6, 0), \alpha]. \tag{7.99}$$

If bf-EVALP clause (7.99) is input to safety property **SP-5** (7.88), then we obtain the EVALP clause,

$$Start(p3, t_3):[(0, 1), \gamma],$$

and the safety for starting process Pr_3 is assured.

7.6   Reasoning in Bf-EVALPSN

In this section, we introduce the process before-after relation reasoning system in bf-EVALPSN, which consists of two inference rules in bf-EVALP. The first basic inference rules for bf-relations according to the before-after relations of process start/finish times, and the second one is the transitive inference rules for bf-relations, which can infer the transitive bf-relation from two continuous bf-relations.

7.6.1   Basic Reasoning for Bf-Relation

We introduce the basic inference rules of bf-relations with referring to Example 5.1 in Sect. 7.5.2, which are called *basic bf-inference rules*. Hereafter we call the inference rules as *ba-inf rules* shortly. First of all, in order to represent the basic bf-inference rules in bf-EVALPSN, we introduce the following literals for expressing process start/finish information again:

$fi(p_i, t)$, which is intuitively interpreted that process Pr_i finishes at time t.

Those literals are used for expressing process finish information and may have one of vector annotations $\bot(0, 0)$, $t(1, 0)$, $f(0, 1)$ and $\top(1, 1)$, where annotations $t(1, 0)$ and $f(0, 1)$ can be intuitively interpreted as "true" and "false", respectively. We show a group of ba-inf rules to be applied at the initial stage (time t_0) for bf-relation reasoning, which are named $(0, 0)$-*rules*.

(0, 0)-rules
Suppose that no process has started yet and the vector annotation of bf-literal $R(p_i, p_j, t)$ is $(0, 0)$, which shows that there is no knowledge in terms of the bf-relation between processes Pr_i and Pr_j, then the following two ba-inf rules can be applied at the initial stage.

$(0, 0)$-*rule-1*   If process Pr_i has started before process Pr_j tarts, then the vector annotation $(0, 0)$ of bf-literal $R(p_i, p_j, t)$ should turn to bf-annotation $be(0, 8)$, which is the greatest lower bound of the set of bf-annotations

$$\{db(0, 12), \ mb(1, 11), \ jb(2, 10), \ sb(3, 9), \ ib(4, 8)\}.$$

$(0, 0)$-*rule-2*   If both processes Pr_i and Pr_j have started at the same time, then it is reasonably anticipated that the bf-relation between processes Pr_i and Pr_j will be in the set of bf-annotations,

$$\{fb(5, 7), \ pba(6, 6), \ fa(7, 5)\}$$

whose greatest lower bound is $(5, 5)$ (refer to Fig. 7.18). Therefore the vector annotation $(0, 0)$ of bf-literal $R(p_i, p_j, t)$ should turn to vector annotation $(5, 5)$.

Ba-inf rules $(0, 0)$-rule-1 and $(0, 0)$-rule-2 may be translated into the bf-EVALPSN clauses,

$$R(p_i, p_j, t):[(0,0), \alpha] \land st(p_i, t):[\mathtt{t}, \alpha] \land \sim st(p_j, t):[\mathtt{t}, \alpha]$$
$$\rightarrow R(p_i, p_j, t):[(0, 8), \alpha], \tag{7.100}$$

$$R(p_i, p_j, t):[(0,0), \alpha] \land st(p_i, t):[\mathtt{t}, \alpha] \land st(p_j, t):[\mathtt{t}, \alpha]$$
$$\rightarrow R(p_i, p_j, t):[(5, 5), \alpha]. \tag{7.101}$$

Suppose that one of ba-inf rules $(0, 0)$-rule-1 and 2 has been applied, then the vector annotation of bf-literal $R(p_i, p_j, t)$ should be one of $(0, 8)$ or $(5, 5)$. Therefore we need to consider two groups of ba-inf rules to be applied after ba-inf rules $(0, 0)$-rule-1 and $(0, 0)$-rule-2, which are named $(0, 8)$-*rules* and $(5, 5)$-*rules*, respectively.
$(0, 8)$-rules
Suppose that process Pr_i has started before process Pr_j starts, then the vector annotation of bf-literal $R(p_i, p_j, t)$ should be $(0, 8)$. We have the following inference rules to be applied after ba-inf rule $(0, 0)$-rule-1.

$(0, 8)$-*rule-1*   If process Pr_i has finished before process Pr_j starts, and process Pr_j has started immediately after process Pr_i finishes, then the vector annotation $(0, 8)$ of bf-literal $R(p_i, p_j, t)$ should turn to bf-annotation $\mathtt{mb}(1, 11)$.

$(0, 8)$-*rule-2*   If process Pr_i has finished before process Pr_j starts, and process Pr_j has not started immediately after process Pr_i has finished, then the vector annotation $(0, 8)$ of bf-literal $R(p_i, p_j, t)$ should turn to bf-annotation $\mathtt{db}(0, 12)$.

$(0, 8)$-*rule-3*   If process Pr_j starts before process Pr_i finishes, then the vector annotation $(0, 8)$ of bf-literal $R(p_i, p_j, t)$ should turn to vector annotation $(2, 8)$ that is the greatest lower bound of the set of bf-annotations,

$$\{\mathtt{jb}(2, 10),\ \mathtt{sb}(3, 9),\ \mathtt{ib}(4, 8)\}.$$

Ba-inf rules $(0, 8)$-rule-1, $(0, 8)$-rule-2 and $(0, 8)$-rule-3 may be translated into the bf-EVALPSN clauses,

$$R(p_i, p_j, t):[(0, 8), \alpha] \land fi(p_i, t):[\mathtt{t}, \alpha] \land st(p_j, t):[\mathtt{t}, \alpha]$$
$$\rightarrow R(p_i, p_j, t):[(1, 11), \alpha], \tag{7.102}$$

$$R(p_i, p_j, t):[(0, 8), \alpha] \land fi(p_i, t):[\mathtt{t}, \alpha] \land \sim st(p_j, t):[\mathtt{t}, \alpha]$$
$$\rightarrow R(p_i, p_j, t):[(0, 12), \alpha], \tag{7.103}$$

$$R(p_i, p_j, t):[(0, 8), \alpha] \land \sim fi(p_i, t):[\mathtt{t}, \alpha] \land st(p_j, t):[\mathtt{t}, \alpha]$$
$$\rightarrow R(p_i, p_j, t):[(2, 8), \alpha]. \tag{7.104}$$

(5, 5)-rules

Suppose that both processes Pr_i and Pr_j have already started at the same time, then the vector annotation of bf-literal $R(p_i, p_j, t)$ should be (5, 5). We have the following inference rules to be applied after ba-inf rule (0, 0)-rule-2.

(5, 5)-*rule-1*   If process Pr_i has finished before process Pr_j finishes, then the vector annotation (5, 5) of bf-literal $R(p_i, p_j, t)$ should turn to bf-annotation sb(5, 7).

(5, 5)-*rule-2*   If both processes Pr_i and Pr_j have finished at the same time, then the vector annotation (5, 5) of bf-literal $R(p_i, p_j, t)$ should turn to bf-annotation pba(6, 6).

(5, 5)-*rule-3*   If process Pr_j has finished before process Pr_i finishes, then the vector annotation (5, 5) of bf-literal $R(p_i, p_j, t)$ should turn to bf-annotation sa(7, 5).

Ba-inf rules (5, 5)-rule-1, (5, 5)-rule-2 and (5, 5)-rule-3 may be translated into the bf-EVALPSN clauses,

$$R(p_i, p_j, t):[(5,5), \alpha] \wedge fi(p_i, t):[t, \alpha] \wedge {\sim}fi(p_j, t):[t, \alpha]$$
$$\rightarrow R(p_i, p_j, t):[(5,7), \alpha], \tag{7.105}$$

$$R(p_i, p_j, t):[(5,5), \alpha] \wedge fi(p_i, t):[t, \alpha] \wedge fi(p_j, t):[t, \alpha]$$
$$\rightarrow R(p_i, p_j, t):[(6,6), \alpha], \tag{7.106}$$

$$R(p_i, p_j, t):[(5,5), \alpha] \wedge {\sim}fi(p_i, t):[t, \alpha] \wedge fi(p_j, t):[t, \alpha]$$
$$\rightarrow R(p_i, p_j, t):[(7,5), \alpha]. \tag{7.107}$$

If ba-inf rules, (0, 8)-rule-1, (0, 8)-rule-2, (5, 5)-rule-1, (5, 5)-rule-2 and (5, 5)-rule-3, and have been applied, bf-relations represented by bf-annotations such as jb(2, 10)/ja(10, 2) between two processes should be derived. On the other hand, even if ba-inf rule (0, 8)-rule-3 has been applied, no bf-annotation could be derived. Therefore a group of ba-inf rules called (2, 8)-rules should be considered after ba-inf rule (0, 8)-rule-3.

(2, 8)-rules

Suppose that process Pr_i has started before process Pr_j starts and process Pr_j has started before process Pr_i finishes, then the vector annotation of bf-literal $R(p_i, p_j, t)$ should be (2, 8) and the following three rules should be considered.

(2, 8)-*rule-1*   If process Pr_i finished before process Pr_j finishes, then the vector annotation (2, 8) of bf-literal $R(p_i, p_j, t)$ should turn to bf-annotation jb(2, 10).

(2, 8)-*rule-2*   If both processes Pr_i and Pr_j have finished at the same time, then the vector annotation (2, 8) of bf-literal $R(p_i, p_j, t)$ should turn to bf-annotation fb(3, 9).

(2, 8)-*rule-3*   If process Pr_j has finished before Pr_i finishes, then the vector annotation (2, 8) of bf-literal $R(p_i, p_j, t)$ should turn to bf-annotation ib(4, 8).

Table 7.2 Application orders of basic bf-inference rules

Vector annotations	Rule	Vector annotation	Rule	Vector annotation	Rule	Vector annotation
$(0, 0)$	Rule-1	$(0, 8)$	Rule-1	$(0, 12)$		
			Rule-2	$(1, 11)$		
			Rule-3	$(2, 8)$	Rule-1	$(2, 10)$
					Rule-2	$(3, 9)$
					Rule-3	$(4, 8)$
	Rule-2	$(5, 5)$	Rule-1	$(5, 7)$		
			Rule-2	$(6, 6)$		
			Rule-3	$(7, 5)$		

Ba-inf rules $(2, 8)$-rule-1, $(2, 8)$-rule-2 and $(2, 8)$-rule-3 may be translated into the bf-EVALPSN clauses,

$$R(p_i, p_j, t) : [(2, 8), \alpha] \wedge fi(p_i, t) : [\mathsf{t}, \alpha] \wedge \sim fi(p_j, t) : [\mathsf{t}, \alpha]$$
$$\rightarrow R(p_i, p_j, t) : [(2, 10), \alpha], \tag{7.108}$$

$$R(p_i, p_j, t) : [(2, 8), \alpha] \wedge fi(p_i, t) : [\mathsf{t}, \alpha] \wedge fi(p_j, t) : [\mathsf{t}, \alpha]$$
$$\rightarrow R(p_i, p_j, t) : [(3, 9), \alpha], \tag{7.109}$$

$$R(p_i, p_j, t) : [(2, 8), \alpha] \wedge \sim fi(p_i, t) : [\mathsf{t}, \alpha] \wedge fi(p_j, t) : [\mathsf{t}, \alpha]$$
$$\rightarrow R(p_i, p_j, t) : [(4, 8), \alpha]. \tag{7.110}$$

The application orders (from the left to the right) of all ba-inf rules are summarized in Table 7.2.

7.6.2 Transitive Reasoning for Bf-Relations

Suppose that a bf-EVALPSN process order control system has to deal with ten processes. Then if it deals with all the bf-relations between ten processes, forty five bf-relations have to be considered. It may take much computing cost. In order to reduce such cost, we consider inference rules to derive the bf-relation between processes Pr_i and Pr_k from two bf-relations between processes Pr_i and Pr_j and between processes Pr_j and Pr_k in bf-EVALPSN, which are named *transitive bf-inference rules*. Hereafter we call transitive bf-inference rules as *tr-inf rules* for short. We introduce how to derive some of tr-inf rules and how to apply them to real-time process order control.

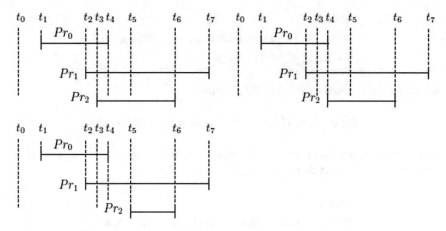

Fig. 7.22 Process time chart 1 (*top left*), 2 (*top right*), 3 (*bottom left*)

Suppose that three processes Pr_0, Pr_1 and Pr_2 are being processed according to the process schedule in Fig. 7.22 in which only the start time of process Pr_2 varies time t_3 to time t_5 and there is no variation of bf-relations between all the processes. The vector annotations of bf-literals $R(p0, p1, t)$, $R(p1, p2, t)$ and $R(p0, p2, t)$ at each time t_i ($i = 1, \ldots, 7$) are shown by the three tables in Table 7.3.

For each table, if we focus on the vector annotations at time t_1 and time t_2, the following tr-inf rule in bf-EVALP clause can be derived:

Table 7.3 Vector annotations of process time chart 1, 2, 3

	t_0	t_1	t_2	t_3	t_4	t_5	t_6	t_7
Process time chart 1								
$R(p0, p1, t)$	(0, 0)	(0, 8)	(2, 8)	(2, 8)	(2, 10)	(2, 10)	(2, 10)	(2, 10)
$R(p1, p2, t)$	(0, 0)	(0, 0)	(0, 8)	(2, 8)	(2, 8)	(2, 8)	(4, 8)	(4, 8)
$R(p0, p2, t)$	(0, 0)	(0, 8)	(0, 8)	(2, 8)	(2, 10)	(2, 10)	(2, 10)	(2, 10)
Process time chart 2								
$R(p0, p1, t)$	(0, 0)	(0, 8)	(2, 8)	(2, 8)	(2, 10)	(2, 10)	(2, 10)	(2, 10)
$R(p1, p2, t)$	(0, 0)	(0, 0)	(0, 8)	(0, 8)	(2, 8)	(2, 8)	(4, 8)	(4, 8)
$R(p0, p2, t)$	(0, 0)	(0, 8)	(0, 8)	(0, 8)	(1, 11)	(1, 11)	(1, 11)	(1, 11)
Process time chart 3								
$R(p0, p1, t)$	(0, 0)	(0, 8)	(2, 8)	(2, 8)	(2, 10)	(2, 10)	(2, 10)	(2, 10)
$R(p1, p2, t)$	(0, 0)	(0, 0)	(0, 8)	(0, 8)	(0, 8)	(2, 8)	(4, 8)	(4, 8)
$R(p0, p2, t)$	(0, 0)	(0, 8)	(0, 8)	(0, 8)	(0, 12)	(0, 12)	(0, 12)	(0, 12)

rule-1

$R(p0, p1, t) : [(0, 8), \alpha] \wedge R(p1, p2, t) : [(0, 0), \alpha]$
$\rightarrow R(p0, p2, t) : [(0, 8), \alpha]$

which may be reduced to the bf-EVALP clause,

$$R(p0, p1, t) : [(0, 8), \alpha] \rightarrow R(p0, p2, t) : [(0, 8), \alpha]. \tag{7.111}$$

Furthermore, if we also focus on the vector annotations at time t_3 and time t_4 in Table 7.3, the following two tr-inf rules also can be derived:

rule-2

$R(p0, p1, t) : [(2, 8), \alpha] \wedge R(p1, p2, t) : [(2, 8), \alpha]$
$$\rightarrow R(p0, p2, t) : [(2, 8), \alpha], \tag{7.112}$$

rule-3

$R(p0, p1, t) : [(2, 10), \alpha] \wedge R(p1, p2, t) : [(2, 8), \alpha]$
$$\rightarrow R(p0, p2, t) : [(2, 10), \alpha]. \tag{7.113}$$

As well as tr-inf rules **rule-2** and **rule-3**, the following two tr-inf rules also can be derived with focusing on the variation of the vector annotations at time t_4.

rule-4

$R(p0, p1, t) : [(2, 10), \alpha] \wedge R(p1, p2, t) : [(2, 8), \alpha]$
$$\rightarrow R(p0, p2, t) : [(1, 11), \alpha], \tag{7.114}$$

rule-5

$R(p0, p1, t) : [(2, 10), \alpha] \wedge R(p1, p2, t) : [(0, 8), \alpha]$
$$\rightarrow R(p0, p2, t) : [(0, 12), \alpha]. \tag{7.115}$$

Among all the tr-inf rules only tr-inf rules **rule-3** and **rule 4** have the same precedent(body),

$$R(p0, p1, t) : [(2, 10), \alpha] \wedge R(p1, p2, t) : [(2, 8), \alpha],$$

and different consequents(heads),

$$R(p0, p2, t) : [(2, 10), \alpha] \quad \text{and} \quad R(p0, p2, t) : [(1, 11), \alpha].$$

Having the same precedent may cause duplicate application of those tr-inf rules. If we take tr-inf rules **rule-3** and **rule-4** into account, obviously they cannot be uniquely applied. In order to avoid duplicate application of tr-inf rules **rule-3** and **rule-4**, we

consider all correct applicable orders **order-1**, **order-2** and **order-3** for all tr-inf rules
rule-1, ..., **rule-5**.

$$\text{order-1:} \quad \text{rule-1} \longrightarrow \text{rule-2} \longrightarrow \text{rule-3} \tag{7.116}$$

$$\text{order-2:} \quad \text{rule-1} \longrightarrow \text{rule-4} \tag{7.117}$$

$$\text{order-3:} \quad \text{rule-1} \longrightarrow \text{rule-5} \tag{7.118}$$

As indicated in the above orders, tr-inf rule **rule 3** should be applied immediately after
tr-inf rule **rule 2**, on the other hand, tr-inf rule **rule 4** should be done immediately after
tr-inf rule **rule 1**. Thus if we take the applicable orders (7.116)–(7.118) into account,
such confusion may be avoidable. Actually, tr-inf rules are not complete, that is to
say there exist some cases in which bf-relations cannot be uniquely determined by
only tr-inf rules.

We show an application of tr-inf rules by taking process time chart 3 in Fig. 7.22
as an example.

At time t_1, tr-inf rule **rule-1** is applied and we have the bf-EVALPSN clause,

$$R(p0, p2, t_1) : [(0, 8), \alpha].$$

At time t_2 and time t_3, no tr-inf rule can be applied and we still have the same vector
annotation $(0, 8)$ of bf-literal $R(p0, p2, t_3)$.
At time t_4, only tr-inf rule **rule-5** can be applied and we obtain the bf-EVALP clause,

$$R(p0, p2, t_4) : [(0, 12), \alpha]$$

and the bf-relation between processes Pr_0 and Pr_2 has been inferred according to
process order **order-3** (7.118).

We could not introduce all tr-inf rules in this subsection though, it is sure that
we have many cases that can reduce bf-relation computing cost in bf-EVALPSN
process order control by using tr-inf rules. In real-time process control systems, such
reduction of computing cost is required and significant in practice.

As another topic, we briefly introduce anticipation of bf-relations in bf-EVALPSN.
For example, suppose that three processes Pr_0, Pr_1 and Pr_2 have started sequentially,
and only process Pr_1 has finished at time t as shown in Fig. 7.23.

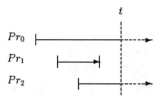

Fig. 7.23 Anticipation of bf-relation

Then, two bf-relations between processes Pr_0 and Pr_1 and between processes Pr_1 and Pr_2 have been already determined, and we have the following two bf-EVALP clauses with the final bf-annotations of the bf-literals,

$$R(p0, p1, t):[\text{ib}(4, 8), \alpha] \quad \text{and} \quad R(p1, p2, t):[\text{mb}(1, 11), \alpha]. \tag{7.119}$$

On the other hand, the bf-relation between processes Pr_0 and Pr_2 cannot be determined yet. However, if we use the tr-inf rule,

rule-6

$$R(p0, p1, t):[(4, 8), \alpha] \wedge R(p1, p2, t):[(2, 10), \alpha]$$
$$\rightarrow R(p0, p2, t):[(2, 8), \alpha], \tag{7.120}$$

we obtain vector annotation $(2, 8)$ as the bf-annotation of bf-literal $R(p0, p2, t)$. Moreover, it is logically anticipated that the bf-relation between processes Pr_0 and Pr_2 will be finally represented by one of three bf-annotations (vector annotations), $\text{jb}(2, 10)$, $\text{sb}(3, 9)$ and $\text{ib}(4, 8)$, since the vector annotation $(2, 8)$ is the greatest lower bound of the set of vector annotations, $\{(2, 10), (3, 9), (4, 8)\}$. As mentioned above, we can systematically anticipate the final bf-annotations from incomplete bf-annotations in bf-EVALPSN. This remarkable anticipatory feature of bf-EVALPSN reasoning could be applied to safety verification and intelligent control that may require such logical anticipation of bf-relations.

7.6.3   Transitive Bf-Inference Rules

In this subsection we list up all transitive bf-inference rules (tr-inf rules) with taking their application orders into account. For simplicity, we represent the tr-inf rule,

$$R(p_i, p_j, t):[(n_1, n_2), \alpha] \wedge R(p_j, p_k, t):[(n_3, n_4), \alpha] \rightarrow R(p_i, p_k, t):[(n_5, n_6), \alpha]$$

by only vector annotations and logical connectives $\wedge$ and $\rightarrow$, as follows:

$$(n_1, n_2) \wedge (n_3, n_4) \rightarrow (n_5, n_6)$$

in the list of tr-inf rules.

Transitive Bf-inference Rules

$$\textbf{TR0} \quad (0, 0) \wedge (0, 0) \rightarrow (0, 0)$$
$$\textbf{TR1} \quad (0, 8) \wedge (0, 0) \rightarrow (0, 8)$$
$$\textbf{TR1} - \textbf{1} \quad (0, 12) \wedge (0, 0) \rightarrow (0, 12)$$
$$\textbf{TR1} - \textbf{2} \quad (1, 11) \wedge (0, 8) \rightarrow (0, 12)$$

$$\textbf{TR1} - \textbf{3} \quad (1, 11) \wedge (5, 5) \rightarrow (1, 11)$$

$$\textbf{TR1} - \textbf{4} \quad (2, 8) \wedge (0, 8) \rightarrow (0, 8)$$

$$\textbf{TR1} - \textbf{4} - \textbf{1} \quad (2, 10) \wedge (0, 8) \rightarrow (0, 12)$$

$$\textbf{TR1} - \textbf{4} - \textbf{2} \quad (4, 8) \wedge (0, 12) \rightarrow (0, 8) \tag{7.121}$$

$$\textbf{TR1} - \textbf{4} - \textbf{3} \quad (2, 8) \wedge (2, 8) \rightarrow (2, 8)$$

$$\textbf{TR1} - \textbf{4} - \textbf{3} - \textbf{1} \quad (2, 10) \wedge (2, 8) \rightarrow (2, 10)$$

$$\textbf{TR1} - \textbf{4} - \textbf{3} - \textbf{2} \quad (4, 8) \wedge (2, 10) \rightarrow (2, 8) \tag{7.122}$$

$$\textbf{TR1} - \textbf{4} - \textbf{3} - \textbf{3} \quad (2, 8) \wedge (4, 8) \rightarrow (4, 8)$$

$$\textbf{TR1} - \textbf{4} - \textbf{3} - \textbf{4} \quad (3, 9) \wedge (2, 10) \rightarrow (2, 10)$$

$$\textbf{TR1} - \textbf{4} - \textbf{3} - \textbf{5} \quad (2, 10) \wedge (4, 8) \rightarrow (3, 9)$$

$$\textbf{TR1} - \textbf{4} - \textbf{3} - \textbf{6} \quad (4, 8) \wedge (3, 9) \rightarrow (4, 8)$$

$$\textbf{TR1} - \textbf{4} - \textbf{3} - \textbf{7} \quad (3, 9) \wedge (3, 9) \rightarrow (3, 9)$$

$$\textbf{TR1} - \textbf{4} - \textbf{4} \quad (3, 9) \wedge (0, 12) \rightarrow (0, 12)$$

$$\textbf{TR1} - \textbf{4} - \textbf{5} \quad (2, 10) \wedge (2, 8) \rightarrow (1, 11)$$

$$\textbf{TR1} - \textbf{4} - \textbf{6} \quad (4, 8) \wedge (1, 11) \rightarrow (2, 8) \tag{7.123}$$

$$\textbf{TR1} - \textbf{4} - \textbf{7} \quad (3, 9) \wedge (1, 11) \rightarrow (1, 11)$$

$$\textbf{TR1} - \textbf{5} \quad (2, 8) \wedge (5, 5) \rightarrow (2, 8)$$

$$\textbf{TR1} - \textbf{5} - \textbf{1} \quad (4, 8) \wedge (5, 7) \rightarrow (2, 8) \tag{7.124}$$

$$\textbf{TR1} - \textbf{5} - \textbf{2} \quad (2, 8) \wedge (7, 5) \rightarrow (4, 8)$$

$$\textbf{TR1} - \textbf{5} - \textbf{3} \quad (3, 9) \wedge (5, 7) \rightarrow (2, 10)$$

$$\textbf{TR1} - \textbf{5} - \textbf{4} \quad (2, 10) \wedge (7, 5) \rightarrow (3, 9)$$

$$\textbf{TR2} \quad (5, 5) \wedge (0, 8) \rightarrow (0, 8)$$

$$\textbf{TR2} - \textbf{1} \quad (5, 7) \wedge (0, 8) \rightarrow (0, 12)$$

$$\textbf{TR2} - \textbf{2} \quad (7, 5) \wedge (0, 12) \rightarrow (0, 8) \tag{7.125}$$

$$\textbf{TR2} - \textbf{3} \quad (5, 5) \wedge (2, 8) \rightarrow (2, 8)$$

$$\textbf{TR2} - \textbf{3} - \textbf{1} \quad (5, 7) \wedge (2, 8) \rightarrow (2, 10)$$

$$\textbf{TR2} - \textbf{3} - \textbf{2} \quad (7, 5) \wedge (2, 10) \rightarrow (2, 8) \tag{7.126}$$

$$\textbf{TR2} - \textbf{3} - \textbf{3} \quad (5, 5) \wedge (4, 8) \rightarrow (4, 8)$$

$$\textbf{TR2} - \textbf{3} - \textbf{4} \quad (7, 5) \wedge (3, 9) \rightarrow (4, 8)$$

$$\textbf{TR2} - \textbf{4} \quad (5, 7) \wedge (2, 8) \rightarrow (1, 11)$$

$$\textbf{TR2} - \textbf{5} \quad (7, 5) \wedge (1, 11) \rightarrow (2, 8) \tag{7.127}$$

$$\textbf{TR3} \quad (5, 5) \wedge (5, 5) \rightarrow (5, 5)$$

$$\textbf{TR3} - \textbf{1} \quad (7, 5) \wedge (5, 7) \rightarrow (5, 5) \tag{7.128}$$

$$\textbf{TR3} - \textbf{2} \quad (5, 7) \wedge (7, 5) \rightarrow (6, 6)$$

Note that the bottom vector annotation $(0, 0)$ in tr-inf rules implies that for any non-negative integers m and n bf-EVALP clause $R(p_j, p_k, t):[(n, m), \alpha]$ satisfies $R(p_j, p_k, t):[(0, 0), \alpha]$ it.

Here we emphasize two important points (I) and (II) in terms of application of tr-inf rules.

(I) Names of tr-inf rules such as TR1-4-3 show their application orders. For example, if tr-inf rule TR1 has been applied, one of tr-inf rules TR1-1,TR1-2,... or TR1-5 should be applied at the following stage; if tr-inf rule TR1-4 has been applied after tr-inf rule TR1, one of tr-inf rules TR1-4-1,TR1-4-2, ... or TR1-4-7 should be applied at the following stage; on the other hand, if one of tr-inf rules TR1-1, TR1-2 or TR1-3 has been applied after tr-inf rule TR1, there is no tr-inf rule to be applied at the following stage because bf-annotations $db(0, 12)$ or $mb(1, 11)$ between processes Pr_i and Pr_k have been already derived.

(II) the following eight tr-inf rules,

TR1-4-2	(121),	TR2-2	(125),
TR1-4-3-2	(122),	TR2-3-2	(126),
TR1-4-6	(123),	TR2-5	(127),
TR1-5-1	(124),	TR3-1	(128)

have no following rule to be applied at the following stage, even though they cannot derive the final bf-relations between processes represented by bf-annotations such as $jb(2, 10)/ja(10, 2)$. For example, suppose that tr-inf rule TR1-4-3-2 has been applied, then the vector annotation $(2, 8)$ of the bf-literal (p_i, p_k, t) just implies that the final bf-relation between processes Pr_i and Pr_k is one of three bf-annotations, $jb(2, 10)$, $sb(3, 9)$ and $ib(4, 8)$. Therefore, if one of the above eight tr-inf rules has been applied, one of ba-inf rules $(0, 8)$-rule, $(2, 8)$-rule or $(5, 5)$-rule should be applied for deriving the final bf-annotation at the following stage. For instance, if tr-inf rule TR1-4-3-2 has been applied, ba-inf rule $(2, 8)$-rule should be applied at the following stage.

Now we show a simple example of bf-relation reasoning by tr-inf rules taking the process time chart 3(bottom left) in Fig. 7.22.

Example 6.1 At time t_1, tr-inf rule TR1 is applied and we have the bf-EVALP clause,

$$R(p_i, p_k, t_1):[(0, 8), \alpha].$$

At time t_2, tr-inf rule TR1-2 is applied, however bf-literal $R(p_i, p_k, t_2)$ has the same vector annotation $(0, 8)$ as the previous time t_1. Therefore we have the bf-EVALP clause,

$$R(p_i, p_k, t_2):[(0, 8), \alpha].$$

At time t_3, no transitive bf-inference rule can be applied, since the vector annotations of bf-literals $R(p_i, p_j, t_3)$ and $R(p_j, p_k, t_3)$ are the same as the previous time t_2. Therefore we still have the bf-EVALP clause having the same vector annotation,

$$R(p_i, p_k, t_3) : [(0, 8), \alpha].$$

At time t_4, tr-inf rule TR1-2-1 is applied and we obtain the bf-EVALP clause having bf-annotation $db(0, 12)$,

$$R(p_i, p_k, t_4) : [(0, 12), \alpha].$$

7.7 Conclusions and Remarks

In this chapter, we have surveyed paraconsistent annotated logic programs EVALPSN and bf-EVALPSN to deal with process before-after relations and have introduced the safety verification method and intelligent process control based on EVALPSN/bf-EVALPSN as applications. The bf-EVALPSN safety verification based process order control method can be applied to various process order control systems requiring real-time processing.

Allen et al. proposed an interval temporal logic for for knowledge representation of properties, actions and events [2, 3]. In the interval temporal logic, predicates such as *Meets(m,n)* are used for representing primitive before-after relations between time intervals m and n, and other before-after relations are represented by six predicates such as *Before*, *Overlaps*, etc. It is well known that the interval temporal logic is a logically sophisticated tool to develop practical planning or natural language understanding systems [2, 3]. However, it does not seem to be so suitable for practical real-time processing because before-after relations between two processes cannot be determined until both of them finish.

On the other hand, in bf-EVALPSN bf-relations are represented more minutely in paraconsistent vector annotations and can be determined according to start/finish information of two processes in real time. Moreover EVALPSN can be implemented on microchips as electronic circuits, although it has not introduced in this chapter. We have already shown that some EVALPSN based control systems can be implemented on a microchips in [29, 38]. Therefore bf-EVALPSN is a more practical tool for dealing with real-time process order control and its safety verification.

In addition to the suitable characteristics for real-time processing, bf-EVALPSN can deal with incomplete and paracomplete knowledge in terms of before-after relation in vector annotations, although the treatment of paracomplete knowledge has not been discussed in this chapter. Furthermore bf-EVALPSN has inference rules for transitive reasoning of before-after relations as shortly described.

Therefore, we can intellectualize various systems by applying EVALPSN and bf-EVALPSN appropriately. We conclude that annotated logic programming is a practical tool for developing an intelligent system.

Acknowledgments We are grateful to the referee and J.M. Abe for useful comments.

References

1. Abe, J.M., Akama, S., Nakamatsu, K.: Introduction to Annotated Logics. Springer, Heidelberg (2015)
2. Allen, J.F.: Towards a general theory of action and time. Artif. Intell. **23**, 123–154 (1984)
3. Allen, J.F., Ferguson, G.: Actions and events in interval temporal logic. J. Logic Comput. **4**, 531–579 (1994)
4. Apt, K.R., Blair, H.A., Walker, A.: Towards a theory of declarative knowledge. In: Minker, J. (ed.) Foundation of Deductive Database and Logic Programs, pp. 89–148. Morgan Kaufmann, CA (1989)
5. Belnap, N.D.: A useful four valued logic. In: Dunn, M., Epstein, G. (eds.) Modern Uses of Multiple-Valued Logic, pp. 8–37. D.Reidel Publishing, Netherlands (1977)
6. Billington, D.: Defeasible logic is stable. J. Logic Comput. **3**, 379–400 (1993)
7. Billington, D.: Conflicting literals and defeasible logic. In: Nayak, A., Pagnucco, M. (eds.) Proceedings of 2nd Australian Workshop Commonsense Reasoning, 1 December, Perth, Australia, Australian Artificial Intelligence Institute, Australia, pp. 1–15 (1997)
8. Blair, H.A., Subrahmanian, V.S.: Paraconsistent logic programming. Theor. Comput. Sci. **68**, 135–154 (1989)
9. da Costa, N.C.A., Subrahmanian, V.S., Vago, C.: The paraconsistent logics $P\mathcal{T}$. Zeitschrift für Mathematische Logic und Grundlangen der Mathematik **37**, 139–148 (1989)
10. Dressler, O.: An extended basic ATMS, In: Reinfrank, M., et al. (eds.) Proceedings of 2nd International Workshop on Non-monotonic Reasoning, 13–15 June, Grassau, Germany, (Lecture Notes in Computer Science LNCS 346), pp. 143–163. Springer, Heidelberg (1988)
11. Fitting, M.: Bilattice and the semantics of logic programming. J. Logic Program. **11**, 91–116 (1991)
12. Gelder, A.V., Ross, K.A, Schlipf, J.S.: The well-founded semantics for general logic programs. J. ACM **38**, 620–650 (1991)
13. Jaskowski, S.: Propositional calculus for contradictory deductive system (English translation of the original Polish paper). Studia Logica **24**, 143–157 (1948)
14. Kifer, M., Subrahmanian, V.S.: Theory of generalized annotated logic programming and its applications. J. Logic Program. **12**, 335–368 (1992)
15. Lloyd, J.W.: Foundations of Logic Programming, 2nd edn. Springer, Berlin (1987)
16. Moore, R.: Semantical considerations on non-monotonic logic. Artif. Intell. **25**, 75–94 (1985)
17. Nakamatsu, K., Suzuki, A.: Annotated semantics for default reasoning. In: Dai, R. (ed.) Proceedings of 3rd Pacific Rim International Conference on Artificial Intelligence (PRICAI94), 15–18 August, Beijin, China, pp. 180–186. International Academic Publishers, China (1994)
18. Nakamatsu, K., Suzuki, A.: A nonmonotonic ATMS based on annotated logic programs. In: Wobcke, W., et al. (eds.) Agents and Multi-Agents Systems (Lecture Notes in Artificial Intelligence LNAI 1441), pp. 79–93. Springer, Berlin (1998)
19. Nakamatsu, K., Abe, J.M.: Reasonings based on vector annotated logic programs. In: Mohammadian, M. (ed.) Computational Intelligence for Modelling, Control and Automation (CIMCA99), (Concurrent Systems Engineering Series 55), pp. 396–403. IOS Press, Netherlands (1999)

20. Nakamatsu, K., Abe, J.M., Suzuki A.: Defeasible reasoning between conflicting agents based on VALPSN. In: Tessier, C., Chaudron, L. (eds.) Proceedings of AAAI Workshop Agents' Conflicts, 18 July, Orland, FL, pp. 20–27. AAAI Press, Menlo Park, CA (1999)
21. Nakamatsu, K., Abe, J.M., Suzuki, A.: Defeasible reasoning based on VALPSN and its application. In: Nayak, A., Pagnucco, M. (eds.) Proceedings of The Third Australian Commonsense Reasoning Workshop, 7 December, Sydney, Australia, University of Newcastle, Sydney, Australia, pp. 114–130 (1999)
22. Nakamatsu, K.: On the relation between vector annotated logic programs and defeasible theories. Logic Log. Philos. **8**, 181–205 (2000)
23. Nakamatsu, K., Abe, J.M., Suzuki, A.: A defeasible deontic reasoning system based on annotated logic programming. In: Dubois, D.M. (ed.) Proceedings of 4th International Conference on Computing Anticipatory Systems (CASYS2000), 7–12 August, 2000, Liege, Belgium, (AIP Conference Proceedings 573), pp. 609–620. American Institute of Physics, NY (2001)
24. Nakamatsu, K., Abe, J.M., Suzuki, A.: Annotated semantics for defeasible deontic reasoning. In: Ziarko, W., Yao, Y. (eds.) Proceedings of 2nd International Conference on Rough Sets and Current Trends in Computing (RSCTC2000), 16–19 October, 2000, Banff, Canada, (Lecture Notes in Artificial Intelligence LNAI 2005), pp. 432–440. Springer, Berlin (2001)
25. Nakamatsu, K., Abe, J.M., Suzuki, A.: Extended vector annotated logic program and its application to robot action control and safety verification. In: Abraham, A., et al. (eds.) Hybrid Information Systems (Advances in Soft Computing Series), pp. 665–680. Physica-Verlag, Heidelberg (2002)
26. Nakamatsu, K., Suito, H., Abe, J.M., Suzuki, A.: Paraconsistent logic program based safety verification for air traffic control, In: El Kamel, A., et al. (eds.) Proceedings of IEEE International Conference on System, Man and Cybernetics 02 (SMC02)
27. Nakamatsu, K., Abe, J.M., Suzuki, A.: A railway interlocking safety verification system based on abductive paraconsistent logic programming. In: Abraham, A., et al. (eds.) Soft Computing Systems (HIS02) (Frontiers in Artificial Intelligence and Applications 87), pp. 775–784. IOS Press, Netherlands (2002)
28. Nakamatsu, K., Abe, J.M., Suzuki, A.: Defeasible deontic robot control based on extended vector annotated logic programming. In: Dubois, D.M. (ed.) Proceedings of 5th International Conference on Computing Anticipatory Systems (CASYS2001) 13–18 August, 2001, Liege, Belgium, (AIP Conference Proceedings 627), American Institute of Physics, New York, NY, pp. 490–500 (2002)
29. Nakamatsu, K., Mita, Y., Shibata, T.: Defeasible deontic action control based on paraconsistent logic program and its hardware application. In: Mohammadian, M. (ed.) Proceedings of International Conference on Computational Intelligence for Modelling Control and Automation 2003(CIMCA2003), 12–14 February, Vienna, Austria, IOS Press, Netherlands (CD-ROM) (2003)
30. Nakamatsu, K., Seno, T., Abe, J.M., Suzuki, A.: Intelligent real-time traffic signal control based on a paraconsistent logic program EVALPSN. In: Wang, G., et al. (eds.) Rough Sets, Fuzzy Sets, Data Mining and Granular Computing (RSFDGrC2003), 26–29 May, Chongqing, China, (Lecture Notes in Artificial Intelligence LNAI 2639), pp. 719–723. Springer, Berlin (2003)
31. Nakamatsu, K., Komaba, H., Suzuki, A.: Defeasible deontic control for discrete events based on EVALPSN. In: Tsumoto, S., et al. (eds.) Proceedings of 4th International Conference on Rough Sets and Current Trends in Computing (RSCTC2004), 1–5 June, Uppsala, Sweeden, (Lecture Notes in Artificial Intelligence LNAI 3066), pp. 310–315. Springer, Berlin (2004)
32. Nakamatsu, K., Ishikawa, R., Suzuki, A.: A paraconsistent based control for a discrete event cat and mouse. In: Negoita, M.G.H., et al. (eds.) Proceedings of 8th International Conference Knowledge-Based Intelligent Information and Engineering Systems (KES2004), 20–25 September, Wellington, NewZealand, (Lecture Notes in Artificial Intelligence LNAI 3214), pp. 954–960. Springer, Berlin (2004)
33. Nakamatsu, K., Chung, S.-L., Komaba, H., Suzuki, A.: A discrete event control based on EVALPSN stable model. In: Slezak, D., et al. (eds.) Rough Sets, Fuzzy Sets, Data Mining and

Granular Computing (RSFDGrC2005), 31 August - 3 September, Regina, Canada, (Lecture Notes in Artificial Intelligence LNAI 3641), pp. 671–681. Springer, Berlin (2005)

34. Nakamatsu, K., Abe, J.M., Akama, S.: An intelligent safety verification based on a paraconsistent logic program. In: Khosla, R., et al. (eds.) Proceedings of 9th International Conference on Knowledge-Based Intelligent Information and Engineering Systems (KES2005), 14–16 September, Melbourne, australia, (Lecture Notes in Artificial Intelligence LNAI 3682), pp. 708–715. Springer, Berlin (2005)

35. Nakamatsu, K., Kawasumi, K., Suzuki, A.: Intelligent verification for pipeline based on EVALPSN. In: Nakamatsu, K., Abe, J.M. (eds.) Advances in Logic Based Intelligent Systems (Frontiers in Artificial Intelligence and Applications 132), pp. 63–70. IOS Press, Netherlands (2005)

36. Nakamatsu, K., Suzuki, A.: Autoepistemic theory and paraconsistent logic program. In: Nakamatsu, K., Abe, J.M. (eds.) Advances in Logic Based Intelligent Systems (Frontiers in Artificial Intelligence and Applications 132), pp. 177–184. IOS Press, Netherlands (2005)

37. Nakamatsu, K., Suzuki, A.: Annotated semantics for non-monotonic reasonings in artificial intelligence—I, II, III, IV. In: Nakamatsu, K., Abe, J.M. (eds.) Advances in Logic Based Intelligent Systems (Frontiers in Artificial Intelligence and Applications 132), pp. 185–215. IOS Press, Netherlands (2005)

38. Nakamatsu, K., Mita, Y., Shibata, T.: An intelligent action control system based on extended vector annotated logic program and its hardware implementation. J. Intell. Autom. Soft Comput. 13, 289–304 (2007)

39. Nakamatsu, K.: Paraconsistent annotated logic program EVALPSN and its application. In: Fulcher, J., Jain, C.L. (eds.) Computational Intelligence: A Compendium (Studies in Computational Intelligence 115), pp. 233–306. Springer, Germany (2008)

40. Nakamatsu, K., Abe, J.M.: The development of paraconsistent annotated logic program. Int. J. Reasoning-Based Intell. Syst. 1, 92–112 (2009)

41. Nakamatsu, K., Abe, J.M., Akama, S.: A logical reasoning system of process before-after relation based on a paraconsistent annotated logic program bf-EVALPSN. KES J. 15(3), 146–163 (2011)

42. Nute, D.: Defeasible reasoning. In: Stohr, E.A., et al. (eds.) Proceedings of 20th Hawaii International Conference on System Science (HICSS87) 1, 6–9 January, Kailua-Kona, Hawaii, University of Hawaii, Hawaii, pp. 470–477 (1987)

43. Nute, D.: Basic defeasible logics. In: Farinas del Cerro, L., Penttonen, M. (eds.) Intensional Logics for Programming, pp. 125–154. Oxford University Press, UK (1992)

44. Nute, D.: Defeasible logic. In: Gabbay, D.M., et al. (eds.) Handbook of Logic in Artificial Intelligence and Logic Programming 3, pp. 353–396. Oxford University Press, UK (1994)

45. Nute, D.: Apparent obligatory. In: Nute, D. (ed.) Defeasible Deontic Logic, (Synthese Library 263), pp. 287–316. Kluwer Academic Publisher, Netherlands (1997)

46. Przymusinski, T.C.: On the declarative semantics of deductive databases and logic programs. In: Minker, J. (ed.) Foundation of Deductive Database and Logic Programs, pp. 193–216. Morgan Kaufmann, New York (1988)

47. Reiter, R.: A logic for default reasoning. Artif. Intell. 13, 81–123 (1980)

48. Shepherdson, J.C.: Negation as failure, completion and stratification. In: Gabbay, D.M., et al. (eds.) Handbook of Logic in Artificial Intelligence and Logic Programming 5, pp. 356–419. Oxford University Press, UK (1998)

49. Subrahmanian, V.S.: Amalgamating knowledge bases. ACM Trans. Database Syst. 19, 291–331 (1994)

50. Subrahmanian, V.S.: On the semantics of qualitative logic programs. In: Proceedings of the 1987 Symposium Logic Programming (SLP87), August 31–September 4, pp. 173–182. IEEE Computer Society Press, CA (1987)

51. Visser, A.: Four valued semantics and the liar. J. Philos. Logic 13, 99–112 (1987)

Chapter 8
A Review on Rough Sets and Possible World Semantics for Modal Logics

Yasuo Kudo, Tetsuya Murai and Seiki Akama

Dedicated to Jair Minoro Abe for his 60th birthday

Abstract It is well known that rough set-based approximations of concepts and possible world semantics of modal logics are closely related. In this chapter, we review the relationships between two types of possible world semantic models, i.e., Kripke model and measure-based model, and two variation of rough sets, i.e., Pawlak's rough set and variable precision rough set.

Keywords Modal logic · Possible world semantics · Kripke model · Measure-based model · Rough set · Variable precision rough set

8.1 Introduction

Rough set theory, proposed by Pawlak [14, 15], provides a theoretical basis of set-based approximations of concepts. Lower and upper approximations by rough set theory are closely related with possible world semantics, i.e., lower approximation and necessity, and upper approximation and possibility. In this chapter, we review the relationships between two types of possible world semantic models, i.e., Kripke

Y. Kudo (✉)
Muroran Institute of Technology, Muroran, Japan
e-mail: kudo@csse.muroran-it.ac.jp

T. Murai
Chitose Institute of Science and Technology, Chitose, Japan
e-mail: t-murai@photon.chitose.ac.jp

S. Akama
C-Republic, 1-20-1 Higashi-Yurigaoka, Asao-ku, Kawasaki 215-0012, Japan
e-mail: akama@jcom.home.ne.jp

© Springer International Publishing Switzerland 2016
S. Akama (ed.), *Towards Paraconsistent Engineering*, Intelligent Systems
Reference Library 110, DOI 10.1007/978-3-319-40418-9_8

model [6] and measure-based model [11, 12], and two variations of rough sets, i.e., Pawlak's rough set [14, 15] and variable precision rough set [26].

The reminder of this chapter is structured as follows. In Sect. 8.2, language and possible world semantics for modal logics are briefly reviewed. In Sect. 8.3, lower and upper approximations in rough sets and variable precision rough sets are discussed. In Sect. 8.4, connections between possible world semantics and rough sets are discussed. Related works about rough set-based semantics for modal logics are mentioned in Sect. 8.5, and finally, we give some conclusion in Sect. 8.6.

8.2 Modal Logics

In this section, we review possible world semantics of modal logics. The contents of this section is mainly based on [2].

8.2.1 Language

Propositional modal logic (for short, modal logic) is an extension of classical propositional logic by adding two unary operators $\Box$ and $\Diamond$, called modal operators, that express the statements $\Box p$ (p is necessary) and $\Diamond p$ (p is possible) for any proposition p.

Suppose $\mathcal{P} = \{p_1, \ldots, p_n(, \ldots)\}$ is a set of finite or countably infinite atomic sentences, $\top$ (truth) and $\bot$ (falsity) are constant sentences, $\wedge$ (conjunction), $\vee$ (disjunction), $\rightarrow$ (conditionality), $\leftrightarrow$ (biconditionality), and $\neg$ (negation) are logical connectives, and $\Box$ (necessity) and $\Diamond$ (possibility) are modal operators. Let $\mathcal{L}_{\mathrm{ML}}(\mathcal{P})$ be the set of sentences of modal logic constructed from the above symbols by the following construction rules:

$$p \in \mathcal{P} \Rightarrow p \in \mathcal{L}_{\mathrm{ML}}(\mathcal{P}), \top, \bot \in \mathcal{L}_{\mathrm{ML}}(\mathcal{P}),$$
$$p \in \mathcal{L}_{\mathrm{ML}}(\mathcal{P}) \Rightarrow \neg p, \Box p, \Diamond p \in \mathcal{L}_{\mathrm{ML}}(\mathcal{P}),$$
$$p, q \in \mathcal{L}_{\mathrm{ML}}(\mathcal{P}) \Rightarrow p \wedge q, p \vee q, p \rightarrow q, p \leftrightarrow q \in \mathcal{L}_{\mathrm{ML}}(\mathcal{P}).$$

We say that a sentence is a modal sentence if the sentence contains at least one modal operator, and otherwise, we say the sentence is a non-modal sentence.

8.2.2 Possible World Semantics for Modal Logics

8.2.2.1 Kripke Model

In this section, we consider possible world semantics to interpret sentences used in modal logic. A *Kripke model*, one of the most popular frameworks of possible world semantics, is the following triple:

$$\mathcal{M} = (U, R, v), \tag{8.1}$$

where U ($\neq \emptyset$) is the set of possible worlds, R is a binary relation on U called an accessibility relation, and $v : \mathcal{P} \times U \to \{0, 1\}$ is a valuation function that assigns a truth value to each atomic sentence $\mathsf{p} \in \mathcal{P}$ at each world $w \in U$. We define that an atomic sentence p is true at a possible world x by the given Kripke model $\mathcal{M}$ if and only if we have $v(\mathsf{p}, x) = 1$. We say that a Kripke model is finite if its set of possible worlds is a finite set.

We denote $\mathcal{M}, x \models p$ to mean that the sentence p is true at the possible world $x \in U$ by the Kripke model $\mathcal{M}$. Otherwise, we denote $\mathcal{M}, x \not\models x$ to mean that p is false at $x \in U$. Similar to classical propositional logic, for any non-modal sentences $p, q \in \mathcal{L}_{ML}(\mathcal{P})$ and any possible world $x \in U$, interpretation of non-modal sentences by the Kripke model $\mathcal{M}$ is defined as follows:

$$\mathcal{M}, x \models \neg p \iff \mathcal{M}, x \not\models p,$$
$$\mathcal{M}, x \models p \wedge q \iff \mathcal{M}, x \models p \text{ and } \mathcal{M}, x \models q,$$
$$\mathcal{M}, x \models p \vee q \iff \mathcal{M}, x \models p \text{ or } \mathcal{M}, x \models q,$$
$$\mathcal{M}, x \models p \to q \iff \mathcal{M}, x \not\models p \text{ or } \mathcal{M}, x \models q,$$
$$\mathcal{M}, x \models p \leftrightarrow q \iff \mathcal{M}, x \models p \to q \text{ and } \mathcal{M}, x \models q \to p.$$

An *accessibility relation* is used to interpret modal sentences by a Kripke model; a modal sentence $\Box p$ is true at a possible world $x \in U$ by a Kripke model $\mathcal{M}$ if and only if p is true at every possible world y that is accessible from x in $\mathcal{M}$. On the other hand, $\Diamond p$ is true at x if and only if there is at least one possible world y that is accessible from x and p is true at y. Formally, interpretation of modal sentences are defined as follows:

$$\mathcal{M}, x \models \Box p \overset{\text{def}}{\iff} \forall y \in U(xRy \Rightarrow \mathcal{M}, y \models p), \tag{8.2}$$

$$\mathcal{M}, x \models \Diamond p \overset{\text{def}}{\iff} \exists y \in U(xRy \text{ and } \mathcal{M}, y \models p). \tag{8.3}$$

For any sentence $p \in \mathcal{L}_{ML}(\mathcal{P})$, the truth set is the set of possible worlds at which p are true by the Kripke model $\mathcal{M}$, and the truth set is defined as follows:

$$\|p\|^{\mathcal{M}} \overset{\text{def}}{=} \{x \in U \mid \mathcal{M}, x \models p\}. \tag{8.4}$$

We say that a sentence p is true in a Kripke model $\mathcal{M}$ if and only if p is true at every possible world in $\mathcal{M}$. We denote $\mathcal{M} \models p$ if p is true in $\mathcal{M}$.

It is well known that various properties of accessible relations correspond to axiom schemas of modal systems (for details, see [2]). Table 8.1 describes the correspondence between axiom schemas in modal systems and properties of accessibility relations in Kripke models. For example, the modal system $S5$ is sound and complete with respect to the class of all Kripke models that the accessibility relations are equivalence relations. The modal system $S5$ consists of all inference rules and axiom

Table 8.1 Correspondence relationship among axiom schima and accessibility relation

Axiom schima		Accessibility relation
Df◇.	$\Diamond p \leftrightarrow \neg\Box\neg p$	(No condition)
M.	$\Box(p \wedge q) \rightarrow (\Box p \wedge \Box q)$	(No condition)
C.	$(\Box p \wedge \Box q) \rightarrow \Box(p \wedge q)$	(No condition)
N.	$\Box\top$	(No condition)
K.	$\Box(p \rightarrow q) \rightarrow (\Box p \rightarrow \Box q)$	(No condition)
D.	$\Box p \rightarrow \Diamond p$	Serial
P.	$\neg\Box\bot$	Serial
T.	$\Box p \rightarrow p$	Reflexive
B.	$p \rightarrow \Box\Diamond p$	Symmetric
4.	$\Box p \rightarrow \Box\Box p$	Transitive
5.	$\Diamond p \rightarrow \Box\Diamond p$	Euclidian

schemas of propositional logic, the axiom schemas **Df◇**, **K**, **T** and **5** in Table 8.1, and the following inference rule:

$$\textbf{RN.} \text{ from } p \text{ infer } \Box p.$$

8.2.2.2 Measure-Based Semantics

Murai et al. [11, 12] introduced *measure-based semantics* of modal logics. In the measure-based semantics, fuzzy measures assigned to each possible worlds are used to interpret modal sentences.

Let U is a non-empty set. A function $\mu : 2^U \rightarrow [0, 1]$ is called a *fuzzy measure* on U if the function μ satisfies the following conditions:

1. $\mu(U) = 1$,
2. $\mu(\emptyset) = 0$, and
3. $\forall X, Y \subseteq U, \ X \subseteq Y \Rightarrow \mu(X) \leq \mu(Y)$.

Formally, a fuzzy measure model $\mathcal{M}_\mu$ is the following triple:

$$\mathcal{M}_\mu = (U, \{\mu_x\}_{x \in U}, v), \tag{8.5}$$

where U is a set of possible worlds, and v is a valuation. $\{\mu_x\}_{x \in U}$ is a class of fuzzy measures μ_x assigned to all possible worlds $x \in U$.

In measure-based semantics of modal logic, each degree $\alpha \in (0, 1]$ of fuzzy measures corresponds to a modal operator $\Box_\alpha$ [11, 12]. In this paper, however, we fix a degree α and consider α-*level fuzzy measure model*.

Similar to the case of Kripke models, $\mathcal{M}_\mu, x \models p$ indicates that the sentence p is true at the possible world $x \in U$ by the α-level fuzzy measure model $\mathcal{M}_\mu$. Interpretation of non-modal sentences is identical to that in Kripke models. On the other hand, to define the truth value of modal sentences at each world $x \in U$ in the α-level fuzzy measure model M_μ, we use the fuzzy measure μ_x assigned to the world x instead of accessibility relations. Interpretation of modal sentences $\Box p$ at a world x is defined as follows:

$$\mathcal{M}_\mu, x \models \Box p \overset{\text{def}}{\Longleftrightarrow} \mu_x\left(\|p\|^{\mathcal{M}_\mu}\right) \geq \alpha, \tag{8.6}$$

where μ_x is the fuzzy measure assigned to x. By this definition, interpretation of modal sentences $\Diamond p$ is obtained by dual fuzzy measures as follows:

$$\mathcal{M}_\mu, x \models \Diamond p \Longleftrightarrow \mu_x^*\left(\|p\|^{\mathcal{M}_\mu}\right) > 1 - \alpha, \tag{8.7}$$

where the dual fuzzy measure μ_x^* of the assigned fuzzy measure μ_x is defined as $\mu_x^*(X) \overset{\text{def}}{=} 1 - \mu_x(X^c)$ for any $X \subseteq U$.

Note that the modal system *EMNP* is sound and complete with respect to the class of all α-level fuzzy measure models [11, 12], where the system *EMNP* consists of all inference rules and axiom schemas of propositional logic and the axiom schemas **Df$\Diamond$**, **M**, **N**, and **P** in Table 8.1 and the following inference rule:

RE. from $p \leftrightarrow q$ infer $\Box p \leftrightarrow \Box q$.

8.3   Rough Sets

8.3.1   Pawlak's Rough Set

In this section, we review theoretical basis of Pawlak's rough set theory, in particular, lower and upper approximation of concepts. The contents of this section is based on [15, 17].

Let U be a non-empty and finite set of objects called the universe of discourse, and E be an equivalence relation on U called an indiscernibility relation. The ordered pair (U, E) is called a *Pawlak approximation space* that is the basis of approximation in rough set theory.

For any element $x \in U$, the *equivalence class* of x with respect to E is defined as follows:

$$[x]_E \overset{\text{def}}{=} \{y \in U \mid xEy\}. \tag{8.8}$$

The equivalence class $[x]_E$ is the set of objects that are not discernible from x with respect to E. The *quotient set* $U/E \overset{\text{def}}{=} \{[x]_E \mid x \in U\}$ provides a partition of U. According to Pawlak [15], any set $X \subseteq U$ represents a concept, and a set of concepts is called knowledge about U. Thus, the quotient set U/E is called E-*basic knowledge about* U [15].

For any set of objects $X \subseteq U$, the *lower approximation* $\underline{E}(X)$ of X and the *upper approximation* $\overline{E}(X)$ of X by the equivalence relation E are defined as follows, respectively:

$$\underline{E}(X) \stackrel{\text{def}}{=} \{x \in U \mid [x]_E \subseteq X\}, \tag{8.9}$$

$$\overline{E}(X) \stackrel{\text{def}}{=} \{x \in U \mid [x]_E \cap X \neq \emptyset\}. \tag{8.10}$$

The lower approximation $\underline{E}(X)$ of X is the set of objects that are certainly included in X. On the other hand, the upper approximation $\overline{E}(X)$ of X is the set of objects that may be included in X.

If we have $\underline{E}(X) = X = \overline{E}(X)$, we say that X is E-definable, and otherwise, if we have $\underline{E}(X) \subset X \subset \overline{E}(X)$, we say that X is E-rough. The concept X is E-definable means that we can denote X correctly by using background knowledge by E. On the other hand, X is E-rough means that we can not denote the concept correctly based on the background knowledge.

8.3.2 Variable Precision Rough Set

Variable precision rough set models (for short, VPRS) proposed by Ziarko [26] is one extension of Pawlak's rough set theory that provides a theoretical basis to treat probabilistic or inconsistent information in the framework of rough sets.

VPRS is based on the majority inclusion relation. Let $X, Y \subseteq U$ be any subsets of U. The majority inclusion relation is defined by the following measure $c(X, Y)$ of the relative degree of misclassification of X with respect to Y:

$$c(X, Y) \stackrel{\text{def}}{=} \begin{cases} 1 - \dfrac{|X \cap Y|}{|X|}, & \text{if } X \neq \emptyset, \\ 0, & \text{otherwise}, \end{cases} \tag{8.11}$$

where $|X|$ represents the cardinality of the set X. It is easily confirmed that $X \subseteq Y$ holds if and only if $c(X, Y) = 0$ holds.

Formally, the majority inclusion relation $\stackrel{\beta}{\subseteq}$ with a fixed *precision* $\beta \in [0, 0.5)$ is defined using the relative degree of misclassification as follows:

$$X \stackrel{\beta}{\subseteq} Y \stackrel{\text{def}}{\Longleftrightarrow} c(X, Y) \leq \beta, \tag{8.12}$$

where the precision β provides the limit of permissible misclassification [26].

Let (U, E) be a Pawlak approximation space, $X \subseteq U$ be any set of objects, and the degree $\beta \in [0, 0.5)$ be a precision. The β-*lower approximation* $\underline{E}_\beta(X)$ and the β-*upper approximation* $\overline{E}_\beta(X)$ of X are defined as follows:

$$\underline{E}_\beta(X) \stackrel{\text{def}}{=} \left\{ x \in U \,\middle|\, [x]_E \stackrel{\beta}{\subseteq} X \right\} \tag{8.13}$$

$$= \left\{ x \in U \,\middle|\, c\left([x]_E, X\right) \le \beta \right\}, \tag{8.14}$$

$$\overline{E}_\beta(X) \stackrel{\text{def}}{=} \left\{ x \in U \,\middle|\, c\left([x]_E, X\right) < 1 - \beta \right\}. \tag{8.15}$$

As mentioned previously, the precision β represents the threshold degree of misclassification of elements in the equivalence class $[x]_E$ to the set X. Thus, in VPRS, misclassification of elements is allowed if the ratio of misclassification is less than β. Note that the β-lower and β-upper approximations with $\beta = 0$ correspond to Pawlak's lower and upper approximations [26].

8.3.3   Properties of Lower and Upper Approximations

Lower and upper approximations of Pawlak's rough set and VPRS satisfy various properties. Table 8.2 represents some properties of β-lower and upper approximations. The symbol "$\checkmark$" appeared in Table 8.2 means, for each property in Table 8.2, whether the property is satisfied in the case of $\beta = 0$ and $0 < \beta < 0.5$, respectively. For example, it is easily observed that the property **C.** $\underline{E}_\beta(X) \cap \underline{E}_\beta(Y) \subseteq \underline{E}_\beta(X \cap Y)$ does not hold in VPRS with the precision $0 < \beta < 0.5$. Note that symbols assigned to properties like **C.** correspond to axiom schemas in modal logic (for detail, see [2]).

Table 8.2 Some properties of β-lower and upper approximations [7]

Property		$\beta = 0$	$0 < \beta < 0.5$
Df◇.	$\overline{E}_\beta(X) = \underline{E}_\beta(X^c)^c$	$\checkmark$	$\checkmark$
M.	$\underline{E}_\beta(X \cap Y) \subseteq \underline{E}_\beta(X) \cap \underline{E}_\beta(Y)$	$\checkmark$	$\checkmark$
C.	$\underline{E}_\beta(X) \cap \underline{E}_\beta(Y) \subseteq \underline{E}_\beta(X \cap Y)$	$\checkmark$	
N.	$\underline{E}_\beta(U) = U$	$\checkmark$	$\checkmark$
K.	$\underline{E}_\beta(X^c \cup Y) \subseteq \left(\underline{E}_\beta(X)^c \cup \underline{E}_\beta(Y) \right)$	$\checkmark$	
D.	$\underline{E}_\beta(X) \subseteq \overline{E}_\beta(X)$	$\checkmark$	$\checkmark$
P.	$\underline{E}_\beta(\emptyset) = \emptyset$	$\checkmark$	$\checkmark$
T.	$\underline{E}_\beta(X) \subseteq X$	$\checkmark$	
B.	$X \subseteq \underline{E}_\beta(\overline{E}_\beta(X))$	$\checkmark$	
4.	$\underline{E}_\beta(X) \subseteq \underline{E}_\beta(\underline{E}_\beta(X))$	$\checkmark$	$\checkmark$
5.	$\overline{E}_\beta(X) \subseteq \underline{E}_\beta(\overline{E}_\beta(X))$	$\checkmark$	$\checkmark$

8.4 Connections Between Rough Sets and Modal Logics

8.4.1 Pawlak Approximation Spaces as Kripke Models

As we reviewed in Sect. 8.3.1, every Pawlak approximation space (U, E) consists of a finite set U of objects and an equivalence relation E on U. Hence, by regarding each object $x \in U$ as a possible world and the equivalenced relation E as an accessibility relation, and by adding a valuation function $v : \mathcal{P} \times U \to \{0, 1\}$, the structure $\mathcal{M} = (U, E, v)$ induced from the Pawlak approximation space (U, E) is regarded as a special case of Kripke model.

For every Kripke model (U, E, v) induced from a Pawlak approximation space (U, E), it is easily confirmed that the truth conditions of modal sentence $\Box p$ by (8.2) is reformulated as follows:

$$\mathcal{M}, x \models \Box p \iff \forall y \in U (xEy \Rightarrow \mathcal{M}, y \models p)$$
$$\iff [x]_E \subseteq \|p\|^{\mathcal{M}} \tag{8.16}$$
$$\iff x \in \underline{E}(\|p\|^{\mathcal{M}}). \tag{8.17}$$

Similarly, the truth condition of modal sentence $\Diamond p$ by (8.3) is also reformulated as follows:

$$\mathcal{M}, x \models \Diamond p \iff \exists y \in U (xEy \text{ and } \mathcal{M}, y \models p)$$
$$\iff [x]_E \cap \|p\|^{\mathcal{M}} \neq \emptyset \tag{8.18}$$
$$\iff x \in \overline{E}(\|p\|^{\mathcal{M}}). \tag{8.19}$$

All axiom schemas in Table 8.1 and the inference rule **RN** are satisfied by every Kripke model with equivalence relation [2], and therefore, the knowledge represented by the Palwak approximation space (U, E) are able to describe by the modal system $S5$.

As a generalization of approximation using rough sets, Yao and Li [21], Yao and Lin [22], and Yao et al. [23] have discussed generalized lower approximation and generalized upper approximation by using arbitrary binary relation R on U instead of the equivalence relation. A pair (U, R) of a finite set U of objects and a binary relation R on U is called an *approximation space*. For every binary relation R on U, a set $U_R(x)$ of objects induced from an object $x \in U$ and R is defined by

$$U_R(x) \overset{\text{def}}{=} \{y \in U \mid xRy\}. \tag{8.20}$$

Obviously, the equivalence class $[x]_E$ by an equivalence relation E is a special case of the set $U_R(x)$. If we regard the set U as the set of possible worlds, the set $U_R(x)$ is the set of accessible possible worlds from the possible world $x \in U$.

For any binary relation R on U and any set $X \subseteq U$, *generalized lower approximation* $\underline{R}(X)$ and *generalized upper approximation* $\overline{R}(X)$ are defined by

$$\underline{R}(X) \stackrel{\text{def}}{=} \{x \in U \mid U_R(x) \subseteq X\}, \tag{8.21}$$

$$\overline{R}(X) \stackrel{\text{def}}{=} \{x \in U \mid U_R(x) \cap X \neq \emptyset\}. \tag{8.22}$$

Similar reformulation of the truth condition of modal operators by (8.17) and (8.19) are also available for the set $U_R(x)$, and therefore, generalized lower and upper approximations of a truth set $\|p\|^{\mathcal{M}}$ correspond to interpretation of modal sentences $\square p$ and $\Diamond p$ in a Kripke model $\mathcal{M} = (U, R, v)$ induced by an approximation space (U, R) with arbitrary binary relation R:

$$\mathcal{M}, x \models \square p \iff U_R(x) \subseteq \|p\|^{\mathcal{M}} \tag{8.23}$$

$$\iff x \in \underline{R}(\|p\|^{\mathcal{M}}), \tag{8.24}$$

$$\mathcal{M}, x \models \Diamond p \iff U_R(x) \cap \|p\|^{\mathcal{M}} \neq \emptyset \tag{8.25}$$

$$\iff x \in \overline{R}(\|p\|^{\mathcal{M}}). \tag{8.26}$$

This fact illustrates close connection between various modal systems and generalized lower and upper approximations, and properties of the binary relation R used for generalized lower and upper approximations correspond to axiom schemas of modal systems as shown in Tables 8.1 and 8.2. Note that Yao [20] also studied theoretical aspects of generalized rough sets induced by arbitrary binary relations.

8.4.2   Possible World Semantics with Variable Precision Rough Sets

Kudo et al. [8] discussed a possible world semantics of modal logics using VPRS by introducing α-level fuzzy measure models based on background knowledge. The original purpose of this model is to provide a unified framework of deduction, induction, and abduction using granularity of possible worlds based on VPRS and measure-based semantics for modal logic.

As we reviewed in previous sections, each equivalence class $[x]_E$ represents a concept and the set of concepts, i.e., the quotient set U/E, describe knowledge by the given Pawak approximation space (U, E). Suppose a Pawlak approximation space (U, E) is given and a Kripke model $\mathcal{M} = (U, E, v)$ induced from (U, E) and a valuation v is considered. In the Kripke model $\mathcal{M}$, any non-modal sentence p that represents a fact is characterized by its truth set $\|p\|^{\mathcal{M}}$. By using the background knowledge, when we consider the fact represented by the non-modal sentence p, we may not need to consider *all* possible worlds in the truth set $\|p\|^{\mathcal{M}}$ and we often consider only *typical situations* about the fact p.

To describe such typical situations of the fact p, the β-lower approximation of the truth set $\|p\|^{\mathcal{M}}$ by the equivalence relation E is examined and regard each possible world in the β-lower approximation of the truth set $\|p\|^{\mathcal{M}}$ as a typical situation about p based on background knowledge U/E. It enables us to regard situations that are not in the β-lower approximation as exceptions of the fact p. Thus, using background knowledge by the quotient set U/E, the following two sets of possible worlds about a fact p are considerable [8]:

- $\|p\|^{\mathcal{M}}$: correct representation of the fact p
- $\underline{E}_{\beta}(\|p\|^{\mathcal{M}})$: the set of typical situations about p (situations that are not typical may also be included)

Using the given Kripke model as background knowledge, an α-level fuzzy measure model to treat typical situations about facts as β-lower approximations in the framework of modal logic are introduced [8]. Let $\mathcal{M} = (U, E, v)$ be a Kripke model induced from a Pawlak approximation space (U, E) and a valuation function $v : \mathcal{P} \times U \to \{0, 1\}$, and $\alpha \in (0.5, 1]$ be a fixed degree. An α-level fuzzy measure model $\mathcal{M}_{\alpha}^{E}$ based on background knowledge is the following triple:

$$\mathcal{M}_{\alpha}^{E} \stackrel{\text{def}}{=} (U, \{\mu_{x}^{E}\}_{x \in U}, v), \tag{8.27}$$

where U and v are the same as in $\mathcal{M}$. The fuzzy measure $\mu_{x}^{E} : 2^{U} \to [0, 1]$ assigned to each $x \in U$ is a *rough membership fucntion* [16], i.e., a probability measure based on the equivalence class $[x]_{E}$ with respect to E, defined by

$$\mu_{x}^{E}(X) \stackrel{\text{def}}{=} \frac{|[x]_{E} \cap X|}{|[x]_{E}|}, \quad \forall X \subseteq U. \tag{8.28}$$

Similar to the case of Kripke-style models, we denote that a sentence p is true at a world $x \in U$ by an α-level fuzzy measure model $\mathcal{M}_{\alpha}^{E}$ by $\mathcal{M}_{\alpha}^{E}, x \models p$. According to the truth valuation of modal sentences in the measure-based semantics by (8.6) and (8.7), truth valuation of modal sentences, $\Box p$ and $\Diamond p$, by the α-level fuzzy measure model $\mathcal{M}_{\alpha}^{E}$ is defined by

$$\mathcal{M}_{\alpha}^{E}, x \models \Box p \stackrel{\text{def}}{\Longleftrightarrow} \mu_{x}^{E}\left(\|p\|^{\mathcal{M}_{\alpha}^{E}}\right) \geq \alpha, \tag{8.29}$$

$$\mathcal{M}_{\alpha}^{E}, x \models \Diamond p \stackrel{\text{def}}{\Longleftrightarrow} \mu_{x}^{E}\left(\|p\|^{\mathcal{M}_{\alpha}^{E}}\right) > 1 - \alpha. \tag{8.30}$$

The truth set of a sentence p in the α-level fuzzy measure model $\mathcal{M}_{\alpha}^{E}$ is defined by

$$\|p\|^{\mathcal{M}_{\alpha}^{E}} \stackrel{\text{def}}{=} \{x \in U \mid \mathcal{M}_{\alpha}^{E}, x \models p\}. \tag{8.31}$$

The constructed α-level fuzzy measure model $\mathcal{M}_{\alpha}^{E}$ from the given Kripke model $\mathcal{M}$ has the following properties.

Theorem 8.1 [8] *Let $\mathcal{M}$ be a finite Kripke model such that its accessibility relation E is an equivalence relation and $\mathcal{M}_\alpha^E$ be the α-level fuzzy measure model based on the background knowledge $\mathcal{M}$ defined by (8.27). For any non-modal sentence $p \in \mathcal{L}_{ML}(\mathcal{P})$ and any sentence $q \in \mathcal{L}_{ML}(\mathcal{P})$, the following equations hold:*

$$\|p\|^{\mathcal{M}_\alpha^E} = \|p\|^{\mathcal{M}}, \tag{8.32}$$

$$\|\Box q\|^{\mathcal{M}_\alpha^E} = \underline{E}_{1-\alpha}\left(\|q\|^{\mathcal{M}_\alpha^E}\right), \tag{8.33}$$

$$\|\Diamond q\|^{\mathcal{M}_\alpha^E} = \overline{E}_{1-\alpha}\left(\|q\|^{\mathcal{M}_\alpha^E}\right). \tag{8.34}$$

Theorem 8.2 (Soundness [8]) *For any α-level fuzzy measure model $\mathcal{M}_\alpha^E$ defined by (8.27) based on any finite Kripke model $\mathcal{M}$ such that its accessibility relation E is an equivalence relation, the following soundness properties are satisfied in the case of $\alpha = 1$ and $\alpha \in (0.5, 1)$, respectively:*

- *If $\alpha = 1$, then all theorems of the system S5 are true in $\mathcal{M}_\alpha^E$.*
- *If $\alpha \in (0.5, 1)$, then all theorems of the system EMND45 are true in $\mathcal{M}_\alpha^E$,*

where the system EMND45 consists of the inference rules and axiom schemas of the system EMNP and the axiom schemas **D**, **4**, *and* **5**.

This result enables us to represent facts and rules in reasoning processes as non-modal sentences and typical situations of facts and rules as lower approximations of truth sets of non-modal sentences [8]. From (8.32) and (8.33) in Theorem 8.1, the α-level fuzzy measure model $\mathcal{M}_\alpha^E$ based on background knowledge $\mathcal{M}$ exhibits the characteristics of correct representations of facts by the truth sets of non-modal sentences and typical situations of the facts by the $(1 - \alpha)$-lower approximations of truth sets of sentences. Thus, a modal sentence $\Box p$ is interpreted as *typically p*, and typical situations are used to characterize semantical aspects of deduction, induction, and abduction in a granularity-based framework [8].

8.5   Related Works

Connections between generalized rough sets and modal logics have been widely discussed with various approaches; Thiele [19] discussed an approach to generalize rough set theory based on arbitrary binary relations and modal logics. Kondo [5] and Zhu [25] discussed some fundamental properties of generalized rough set induced by binary relations. Järvinen et al. [4] discussed connections among modal logic, rough set, and Galois connection. Liau [9, 10] discussed modal logics semantics with probabilistic approximation spaces.

Various kinds of rough-set-based modal logics have also been introduced (e.g. [13]). As one example, Balbiani et al. [1] introduced a modal logic for Pawlak's approximation space with rough cardinality n.

8.6   Conclusion

In this chapter, we reviewed close relationships between rough set-based lower and upper approximations of concepts and possible world semantics of modal logics. We concentrated the relationships between two types of possible world semantic models, i.e., Kripke model and measure-based model, and two types of rough sets, i.e., Pawlak's rough set and VPRS. Relationships between possible world semantics and other various types of rough sets, i.e., covering-based rough set [24], dominance-based rough set [3], and Bayesian rough set [18], will be explored in future issues.

References

1. Balbiani, P., Iliev, P., Vakarelov, D.: A modal logic for pawlaks approximation spaces with rough cardinality n. Fundamenta Informaticae, **83**, 451E464 (2008)
2. Chellas, B.F.: Modal Logic: An Introduction. Cambridge University Press (1980)
3. Greco, S., Matarazzo, B., Słowiński, R.: Rough set theory for multicriteria decision analysis. Eur. J. Oper. Res. **129**, 1–47 (2002)
4. Järvinen, J., Kondo, M., Kortelainen, J.: Logics from Galois connections. Int. J. Approximate Reasoning **49**, 595 E606 (2008)
5. Kondo, M.: On the structure of generalized rough sets. Inf. Sci. **176**, 589–E00 (2006)
6. Kripke, S.A.: Semantical analysis of modal logic I. Normal modal propositional calculi. Zeitschr. 1. math. Logik und Otundlagen d. Math. **9**, 67–96 (1963)
7. Kudo, Y., Murai, T.: Approximation of concepts and reasoning based on rough sets. J. Japn. Soc. Artif. Intell. **22**(5), 597–604 (2007) (in Japanese)
8. Kudo, Y., Murai, T., Akama, S.: A granularity-based framework of deduction, induction, and abduction. Int. J. Approximate Reasoning **50**, 1215–1226 (2009)
9. Liau, C.J.: An overview of rough set semantics for modal and quantifier logics. Int. J. Uncertainty Fuzziness Knowl. Based Syst. **8**(1), 93–118 (2000)
10. Liau, C.J.: Modal reasoning and rough set theory. In: Artificial Intelligence: Methodology, Systems, and Applications, LNCS, vol. 1480, pp. 317–30. Springer (2006)
11. Murai, T., Miyakoshi, M., Shimbo, M.: Measure-Based Semantics for Modal Logic. In: Fuzzy Logic : State of the Art, pp. 395–405. Kluwer (1993)
12. Murai, T., Miyakoshi, M., Shimbo, M.: A logical foundation of graded modal operators defined by fuzzy measures. In: Proceedings of 4th FUZZ-IEEE, pp. 151–156 (1995)
13. Orłowska, E. (ed.): Incomplete Information: Rough Set Analysis. Physica-Verlag, Springer (1998)
14. Pawlak, Z.: Rough Sets. Int. J. Comput. Inf. Sci. **11**, 341–356 (1982)
15. Pawlak, Z.: Rough Sets: Theoretical Aspects of Reasoning about Data. Kluwer (1991)
16. Pawlak, Z., Skowron, A.: Rough membership functions: a tool for reasoning with uncertainty. In: Algebraic Methods in Logic and In Computer Science, vol. 28. Banach Center Publications, Institute of Mathematics, Polish Academy of Sciences, Warszawa (1993)
17. Polkowski, L.: Rough sets: Mathematical Foundations. Advances in Soft Computing. Physica-Verlag (2002)
18. Ślęzak, D., Ziarko, W.: The investigation of the Bayesian rough set model. Int. J. Approximate Reasoning **40**, 81–91 (2005)
19. Thiele, H.: Generalizing the explicit concept of rough set on the basis of modal logic. In: Reusch, B., et al. (eds.) Computational Intelligence in Theory and Practice. Springer, Berlin (2001)

20. Yao, Y.Y.: Generalized rough set models. In: Polkowski, L., Skowron, A. (eds.) Rough Sets in Knowledge Discovery, pp. 286–318. Physica-Verlag, Heidelberg (1998)
21. Yao, Y.Y., Li, X.: Comparison of rough-set and interval-set models for uncertain reasoning. Fundamenta Informaticae **27**(2–3), 289–298 (1996)
22. Yao, Y.Y., Lin, T.Y.: Generalization of rough sets using modal logics. Intell. Autom. Soft Comput. **2**(2), 103–120 (1996)
23. Yao, Y.Y., Wang, S.K.M., Lin, T.Y.: A review of rough set models. In: Rough Sets and Data Mining, pp. 47–75. Kluwer (1997)
24. Zakowski, W.: Approximations in the space (u, π). Demonstratio Mathematica **16**, 761–769 (1983)
25. Zhu, W.: Generalized rough sets based on relations. Inf. Sci. **177**, 4997E5011 (2007)
26. Ziarko, W.: Variable precision rough set model. J. Comput. Syst. Sci. **46**, 39–59 (1993)

Chapter 9
Paraconsistency, Chellas's Conditional Logics, and Association Rules

Tetsuya Murai, Yasuo Kudo and Seiki Akama

Dedicated to Jair Minoro Abe for his 60th birthday

Abstract Paraconsistency and its dual paracompleteness are now counted as key concepts in intelligent decision systems because so much inconsistent and incomplete information can be found around us. In this paper, a framework of conditional models for conditional logic and their measure-based extensions are introduced in order to represent association rules in a logical way. Then paracomplete and paraconsistent aspects of conditionals are examined in the framework. Finally we apply conditionals into the definition of association rules in data mining with confidence and consider their extension to the case of Dempster-Shaer theory of evidence serving double-indexed confidence.

Keywords Paraconsistency · Paracompleteness · Conditional logics · Measure-based semantics · Association rules

T. Murai (✉)
Chitose Institute of Science and Technology, Chitose 066-8655, Japan
e-mail: t-murai@photon.chitose.ac.jp

Y. Kudo
Muroran Institute of Technology, Muroran, Japan
e-mail: kudo@csse.muroran-it.ac.jp

S. Akama
C-Republic, 1-20-1 Higashi-Yurigaoka, Asao-ku, Kawasaki 215-0012, Japan
e-mail: akama@jcom.home.ne.jp

© Springer International Publishing Switzerland 2016
S. Akama (ed.), *Towards Paraconsistent Engineering*, Intelligent Systems
Reference Library 110, DOI 10.1007/978-3-319-40418-9_9

9.1 Introduction

The authors have tried to give a kind of logical foundation to data mining. Murai et al. [15–17] tried to present a logical formulation of association rules [1–3] using Chellas's conditional models for conditional logics [7] and their measure-based extensions (cf. [12–14]). Akama and Abe [6] proposed a comprehensive idea of paraconsistent logic databases as data warehouse based on paraconsistent and annotated logics [4, 5, 8].

In our opinion, paraconsistency and its dual paracompleteness become key concepts in future development of intelligent decision systems because nowadays there are so much inconsistent and incomplete information around us. In classical logic, inconsistency means triviality in the sense that all sentences become theorems. Paraconsistency means inconsistency but non-triviality. Thus we need new kinds of logic like paraconsistent and annotated logics [4, 5, 8]. Paracompleteness is the dual concept of paraconsistency where the excluded middle is not true.

In this paper, we put association rules in a framework of conditional models [7] and their measure-based extensions (cf. [12–14]) and examine their paracomplete and paraconsistent aspects in the framework.

Then we notice that the standard confidence [1] is nothing but a conditional probability where even weights are a priori assigned to each transaction that contains the items in question at the same time. All of such transactions, however, do not necessarily give us such evidence because some co-occurrences might be contingent. For describing such cases we further introduce double-indexed confidence based on Dempster-Shafer theory [19].

9.2 Chellas's Conditional Models and Their Measure-Based Extensions for Conditional Logics

9.2.1 Standard and Minimal Conditional Models

Given a finite set $\mathcal{P}$ of items as *atomic sentences*, a *language* $\mathcal{L}_{\mathrm{CL}}(\mathcal{P})$ for conditional logic is formed from $\mathcal{P}$ as the set of sentences closed under the usual propositional operators such as $\top, \bot, \neg, \wedge, \vee, \rightarrow$, and $\leftrightarrow$ as well as $\Box\!\!\rightarrow$ and $\Diamond\!\!\rightarrow$[1] (*two kinds of conditionals*) in the following usual way.

1. If $x \in \mathcal{P}$ then $x \in \mathcal{L}_{\mathrm{CL}}(\mathcal{P})$.
2. $\top, \bot \in \mathcal{L}_{\mathrm{CL}}(\mathcal{P})$.
3. If $p \in \mathcal{L}_{\mathrm{CL}}(\mathcal{P})$ then $\neg p \in \mathcal{L}_{\mathrm{CL}}(\mathcal{P})$.
4. If $p, q \in \mathcal{L}_{\mathrm{CL}}(\mathcal{P})$ then $p \wedge q, p \vee q, p \rightarrow q, p \leftrightarrow q, p\Box\!\!\rightarrow q, p\Diamond\!\!\rightarrow q \in \mathcal{L}_{\mathrm{CL}}(\mathcal{P})$.

---

[1]In [7], Chellas used only $\Box\!\!\rightarrow$. The latter connective $\Diamond\!\!\rightarrow$ follows Lewis [11].

Chellas [7] introduces two kind of models called standard and minimal. There relationship is similar to Kripke and Scott-Montague models for modal logics.

Definition 9.1 (*Chellas* [7], *p. 268*) A *standard conditional model* $\mathcal{M}_{CL}$ for conditional logic is a structure

$$\langle W, f, v \rangle,$$

where W is a non-empty set of possible worlds, v is a truth-assignment function

$$v : \mathcal{P} \times W \to \{0, 1\},$$

and f is a function

$$f : W \times 2^W \to 2^W. \blacksquare$$

The truth conditions for $\Box\!\!\rightarrow$ and $\Diamond\!\!\rightarrow$ in standard conditional models are given by

1. $\mathcal{M}_{CL}, w \models p\Box\!\!\rightarrow q \stackrel{\text{def}}{\Longleftrightarrow} f(w, \|p\|^{\mathcal{M}_{CL}}) \subseteq \|q\|^{\mathcal{M}_{CL}}$,
2. $\mathcal{M}_{CL}, w \models p\Diamond\!\!\rightarrow q \stackrel{\text{def}}{\Longleftrightarrow} f(w, \|p\|^{\mathcal{M}_{CL}}) \cap \|q\|^{\mathcal{M}_{CL}} \neq \emptyset$,

where $\|p\|^{\mathcal{M}_{CL}} = \{w \in W \mid \mathcal{M}_{CL}, w \models p\}$. Thus we have the following relationship between the two kind conditionals:

$$p\Box\!\!\rightarrow q \leftrightarrow \neg(p\Diamond\!\!\rightarrow \neg q).$$

The function f can be regarded as a kind of selection function. That is, $p\Box\!\!\rightarrow q$ is true at a world w when q is true at any world selected by f with respect p and w. Similarly, $p\Diamond\!\!\rightarrow q$ is true at a world w when q is true at least at one of the worlds selected by f with respect p and w.

A *minimal conditional models* is a Scott-Montague-like extension of standard conditional model [7].

Definition 9.2 (*Chellas* [7], *p. 270*) A *minimal conditional model* $\mathcal{M}_{CL}$ for conditional logic is a structure

$$\langle W, g, v \rangle,$$

where W and v are the same ones as in the standard conditional models. The difference is the second term

$$g : W \times 2^W \to 2^{2^W}. \blacksquare$$

The truth conditions for $\Box\!\!\rightarrow$ and $\Diamond\!\!\rightarrow$ in a minimal conditional model are given by

1. $\mathcal{M}_{CL}, w \models p\Box\!\!\rightarrow q \stackrel{\text{def}}{\Longleftrightarrow} \|q\|^{\mathcal{M}_{CL}} \in g(w, \|p\|^{\mathcal{M}_{CL}})$,
2. $\mathcal{M}_{CL}, w \models p\Diamond\!\!\rightarrow q \stackrel{\text{def}}{\Longleftrightarrow} (\|q\|^{\mathcal{M}_{CL}})^C \notin g(w, \|p\|^{\mathcal{M}_{CL}})$,

Thus we have also the following relationship:

$$p\Box\!\!\rightarrow q \leftrightarrow \neg(p\Diamond\!\!\rightarrow \neg q).$$

Note that, if the function g satisfies the following condition

$$X \in g(w, \|p\|^{\mathcal{M}_{CL}}) \Leftrightarrow \cap g(w, \|p\|^{\mathcal{M}_{CL}}) \subseteq X$$

for every world w and every sentence p, then, by defining

$$f_g(w, \|p\|^{\mathcal{M}_{CL}}) \overset{\text{def}}{=} \cap g(w, \|p\|^{\mathcal{M}_{CL}}),$$

we have the standard conditional model $\langle W, f_g, v \rangle$ that is equivalent to the original minimal model.

9.2.2 Measure-Based Extensions

Next we introduce measure-based extensions of the previous minimal conditional models. Such extensions are models for graded conditional logics.

Given a finite set $\mathcal{P}$ of items as *atomic sentences*, a *language* $\mathcal{L}_{\text{gCL}}(\mathcal{P})$ for graded conditional logic is formed from $\mathcal{P}$ as the set of sentences closed under the usual propositional operators such as $\top$, $\bot$, $\neg$, $\wedge$, $\vee$, $\rightarrow$, and $\leftrightarrow$ as well as $\Box\!\!\!\rightarrow_k$ and $\Diamond\!\!\!\rightarrow_k$ (*graded conditionals*) for $0 < k \leq 1$ in the usual way.

1. If $x \in \mathcal{P}$ then $x \in \mathcal{L}_{\text{gCL}}(\mathcal{P})$.
2. $\top, \bot \in \mathcal{L}_{\text{gCL}}(\mathcal{P})$.
3. If $p \in \mathcal{L}_{\text{gCL}}(\mathcal{P})$ then $\neg p \in \mathcal{L}_{\text{gCL}}(\mathcal{P})$.
4. If $p, q \in \mathcal{L}_{\text{gCL}}(\mathcal{P})$ then $p \wedge q, p \vee q, p \rightarrow q, p \leftrightarrow q \in \mathcal{L}_{\text{gCL}}(\mathcal{P})$,
5. If $[p, q \in \mathcal{L}_{\text{gCL}}(\mathcal{P})$ and $0 < k \leq 1]$ then $p\Box\!\!\!\rightarrow_k q, p\Diamond\!\!\!\rightarrow_k q \in \mathcal{L}_{\text{gCL}}(\mathcal{P})$.

A graded conditional model is defined as a family of minimal conditional model (cf. Chellas [7]):

Definition 9.3 Given a fuzzy measure

$$m : 2^W \times 2^W \rightarrow [0, 1],$$

a *measure-based conditional model* $\mathcal{M}_{\text{gCL}}^m$ for graded conditional logic is a structure

$$\langle W, \{g_k\}_{0<k\leq 1}, v \rangle,$$

where W and V are the same ones as in the standard conditional models. g_k is defined by a fuzzy measure m as

$$g_k(t, X) \overset{\text{def}}{=} \{Y \subseteq 2^W \mid m(Y, X) \geq k\}. \quad \blacksquare$$

The model $\mathcal{M}_{\mathrm{gCL}}^{m}$ is called *finite* because so is W. Further, in this paper, we call the model $\mathcal{M}_{\mathrm{gCL}}^{m}$ *uniform* since functions $\{g_k\}$ in the model does not depend on any world in $\mathcal{M}_{\mathrm{gCL}}^{m}$.

The truth conditions for $\square\!\!\rightarrow_k$ and $\diamond\!\!\rightarrow_k$ in a measure-based conditional model are given by

$$\mathcal{M}_{\mathrm{gCL}}^{m}, t \models p\square\!\!\rightarrow_k q \text{ iff } \|q\|^{\mathcal{M}_{\mathrm{gCL}}^{m}} \in g_k(t, \|p\|^{\mathcal{M}_{\mathrm{gCL}}^{m}}),$$
$$\mathcal{M}_{\mathrm{gCL}}^{m}, t \models p\diamond\!\!\rightarrow_k q \text{ iff } (\|q\|^{\mathcal{M}_{\mathrm{gCL}}^{m}})^{C} \notin g_k(t, \|p\|^{\mathcal{M}_{\mathrm{gCL}}^{m}}).$$

The basic idea of these definitions is the same as in fuzzy-measure-based semantics for graded modal logic defined in [12–14].

When we take m as a conditional probability, the truth conditions of graded conditional becomes

$$\mathcal{M}_{\mathrm{gCL}}^{\mathrm{Pr}}, t \models p\square\!\!\rightarrow_k q \text{ iff } \mathrm{Pr}(\|q\|^{\mathcal{M}_{\mathrm{gCL}}^{\mathrm{Pr}}} \mid \|p\|^{\mathcal{M}_{\mathrm{gCL}}^{\mathrm{Pr}}}) \geq k.$$

We have several soundness results based on probability-measure-based semantics (cf. [12–14]) shown in Table 9.1.

Table 9.1 Soundness results of graded conditionals by probability measures

$0 < k \le \frac{1}{2}$	$\frac{1}{2} < k < 1$	$k = 1$	Rules and axiom schemata
○	○	○	**RCEA.** $\dfrac{p\leftrightarrow q}{(p\square\!\!\rightarrow_k q)\leftrightarrow(q\square\!\!\rightarrow_k q)}$
○	○	○	**RCEC.** $\dfrac{q\leftrightarrow q'}{(p\square\!\!\rightarrow_k q)\leftrightarrow(p\square\!\!\rightarrow_k q')}$
○	○	○	**RCM.** $\dfrac{q\rightarrow q'}{(p\square\!\!\rightarrow_k q)\rightarrow(p\square\!\!\rightarrow_k q')}$
		○	**RCR.** $\dfrac{(q\wedge q')\rightarrow r}{((p\square\!\!\rightarrow_k q)\wedge(p\square\!\!\rightarrow_k q'))\rightarrow(p\square\!\!\rightarrow_k r)}$
○	○	○	**RCN.** $\dfrac{q}{p\square\!\!\rightarrow_k q}$
		○	**RCK.** $\dfrac{(q_1\wedge\cdots\wedge q_n)\rightarrow q}{((p\square\!\!\rightarrow_k q_1)\wedge\cdots\wedge(p\square\!\!\rightarrow_k q_n))\rightarrow(p\square\!\!\rightarrow_k q)}$
○	○	○	**CM.** $(p\square\!\!\rightarrow_k(q\wedge r))\rightarrow(p\square\!\!\rightarrow_k q)\wedge(p\square\!\!\rightarrow_k r)$
		○	**CC.** $(p\square\!\!\rightarrow_k q)\wedge(p\square\!\!\rightarrow_k r)\rightarrow(p\square\!\!\rightarrow_k(q\wedge r))$
		○	**CR.** $(p\square\!\!\rightarrow_k(q\wedge r))\leftrightarrow(p\square\!\!\rightarrow_k q)\wedge(p\square\!\!\rightarrow_k r)$
○	○	○	**CN.** $p\square\!\!\rightarrow_k \top$
○	○	○	**CP.** $\neg(p\square\!\!\rightarrow_k \bot)$
		○	**CK.** $(p\square\!\!\rightarrow_k(q\rightarrow r))\rightarrow(p\square\!\!\rightarrow_k q)\rightarrow(p\square\!\!\rightarrow_k r)$
	○	○	**CD.** $\neg((p\square\!\!\rightarrow_k q)\wedge(p\square\!\!\rightarrow_k \neg q))$
○			**CD$_C$.** $(p\square\!\!\rightarrow_k q)\vee(p\square\!\!\rightarrow_k \neg q)$

9.3 Paraconsistency and Paracompleteness in Conditionals

As Chellas pointed out in his book [7] (p. 269), conditionals $p\Box\!\!\rightarrow q$ (and also $p\Diamond\!\!\rightarrow q$) is regarded as relative modal sentences like $[p]q$ (and also $\langle p\rangle q$). So we first see paraconsistency and paracompleteness in the usual modal setting for convenience.

9.3.1 Modal Logic Case

Let us define some standard language $\mathcal{L}$ for modal logic with two modal operators $\Box$ and $\Diamond$. In [18], we examined some relationship between modal logics and paraconsistency and paracompleteness. Let us assume a language $\mathcal{L}$ of modal logic as usual. In terms of modal logic, paracompleteness and paraconsistency have a close relation to the following axiom schemata:

$$\mathbf{D.} \quad \Box p \rightarrow \neg\Box\neg p,$$
$$\mathbf{D_C.} \quad \neg\Box\neg p \rightarrow \Box p,$$

because they have their equivalent expressions

$$\neg(\Box p \wedge \Box\neg p),$$
$$\Box p \vee \Box\neg p,$$

respectively. That is, given a system of modal logic Σ, define the following set of sentences

$$T \stackrel{\text{def}}{=} \{p \in \mathcal{L} \mid \vdash_\Sigma \Box p\},$$

where $\vdash_\Sigma \Box p$ means $\Box p$ is a theorem of Σ. Then the above two schemata mean that, for any sentence p

$$\text{not}(p \in T \text{ and } \neg p \in T)$$
$$p \in T \text{ or } \neg p \in T$$

respectively, and obviously the former describes the consistency of T and the latter the completeness of T. Thus

- T is inconsistent when Σ does not contain $\mathbf{D}$.
- T is incomplete when Σ does not contain $\mathbf{D_C}$.

A system Σ is regular when it contains the following rule and axiom schemata

$$p \leftrightarrow q \Rightarrow \Box p \leftrightarrow \Box q$$
$$(\Box p \wedge \Box q) \leftrightarrow \Box(p \wedge q)$$

Note that any normal system is regular.

In [18] we pointed out the followings. If Σ is regular, then we have

$$(\Box p \wedge \Box \neg p) \leftrightarrow \Box \neg \top \tag{9.1}$$

where $\bot \leftrightarrow \neg \top$ and $\bot$ is falsity constant, which means inconsistency itself. Thus we have triviality:

$$T = \mathcal{L}.$$

But if Σ is not regular, then we have no longer (9.1), thus, in general

$$T \neq \mathcal{L},$$

which means T is paraconsistent. That is, local inconsistency does not generate triviality as global inconsistency.

9.3.2 Conditional Logic Case

Next we apply the previous idea into conditional logics. In conditional logics, the corresponding axiom schemata

CD. $\neg((p\Box\!\!\rightarrow q) \wedge (p\Box\!\!\rightarrow \neg q))$
CD$_C$. $(p\Box\!\!\rightarrow q) \vee (p\Box\!\!\rightarrow \neg q)$

Given a system CL of conditional logic, define the following set of conditionals (rules):

$$R \stackrel{\text{def}}{=} \{p\Box\!\!\rightarrow q \in \mathcal{L}_{CD} \mid \vdash_{CL} p\Box\!\!\rightarrow q\}.$$

where $\mathcal{L}_{CD}$ is a language for conditional logic and $\vdash_{CL} p\Box\!\!\rightarrow q$ means $p\Box\!\!\rightarrow q$ is a theorem of CL. Then the above two schemata mean that, for any sentence p

$$\text{not}(p\Box\!\!\rightarrow q \in R \text{ and } p\Box\!\!\rightarrow \neg q \in R)$$
$$p\Box\!\!\rightarrow q \in R \text{ or } p\Box\!\!\rightarrow \neg q \in R$$

respectively, and obviously the former describes the consistency of R and the latter the completeness of R. Thus, for the set R of conditionals (rules)

- R is inconsistent when CL does not contain **CD**.
- R is incomplete when CL does not contain **CD$_C$**.

9.4 Paraconsistency and Paracompleteness in Association Rules

9.4.1 Association Rules

Let $\mathcal{I}$ be a finite set of *items*. Any subset X in $\mathcal{I}$ is called an *itemset* in $\mathcal{I}$. A database is comprised of *transactions*, which are actually obtained or observed itemsets. Formally, we give the following definition:

Definition 9.4   A *database* $\mathcal{D}$ on $\mathcal{I}$ is defined as

$$\langle T, V \rangle,$$

where

1. $T = \{1, 2, \ldots, n\}$ (n is the size of the database),
2. $V : T \to 2^{\mathcal{I}}$. ∎

Thus, for each transaction $i \in T$, V gives its corresponding set of items $V(i) \subseteq \mathcal{I}$.
For an itemset X, its *degree of support* $s(X)$ is defined by

$$s(X) \stackrel{\text{def}}{=} \frac{|\{t \in T \mid X \subseteq V(t)\}|}{|T|},$$

where $| \cdot |$ is a size of a finite set.

Definition 9.5 (*Agrawal et al.* [1]) Given a set of items $\mathcal{I}$ and a database $\mathcal{D}$ on $\mathcal{I}$, an *association rule* is an implication of the form

$$X \Longrightarrow Y,$$

where X and Y are itemsets in $\mathcal{I}$ with $X \cap Y = \emptyset$. ∎

The following two indices were introduced in [1].

Definition 9.6 (*Agrawal et al.* [1])

1. An association rule $r = (X \Longrightarrow Y)$ holds with *confidence* $c(r)$ $(0 \le c(r) \le 1)$ in $\mathcal{D}$ if and only if

$$c(r) = \frac{s(X \cup Y)}{s(X)}.$$

2. An association rule $r = (X \Longrightarrow Y)$ has a *degree of support* $s(r)$ $(0 \le s(r) \le 1)$ in $\mathcal{D}$ if and only if

$$s(r) = s(X \cup Y). \quad ∎$$

In this paper, we will deal with the former index.

Mining of association rules is actually performed by generating all rules that have certain minimum support (denoted *minsup*) and minimum confidence (denoted *minconf*) that a user specifies. See, e.g., [1–3] for details of such algorithms for finding association rules.

For example, consider the movie database in Table 9.2, where AH and HM means Ms. Audrey Hepburn and Mr. Henry Mancini, respectively. If you have watched several (famous) Ms. Hepburn's movies, you might hear some wonderful music composed by Mr. Mancini. This can be represented by the association rule

$$r = \{AH\} \Longrightarrow \{HM\}$$

with its confidence

$$c(r) = \frac{s(\{AH\} \cup \{HM\})}{s(\{AH\})} = 0.5$$

and its degree of support

$$s(r) = \frac{|\{T \mid T \subseteq \{AH\} \cup \{HM\}\}|}{|\mathcal{D}|} = \frac{4}{100} = 0.04.$$

Table 9.2 Movie database

No.	Transaction (movie)	AH	HM
1	Secret people	1	
2	Monte Carlo baby	1	
3	Roman holiday	1	
4	My fair lady	1	
5	Breakfast at Tiffany's	1	1
6	Charade	1	1
7	Two for the road	1	1
8	Wait until dark	1	1
9	Days of wine and rose		1
10	The great race		1
11	The pink panther		1
12	Sunflower		1
13	Some like it hot		
14	12 Angry men		
15	The apartment		
	. . . . . .		
100	Les aventuriers		

9.4.2 Measure-Based Conditional Models for Databases

Let us regards a finite set $\mathcal{I}$ of items as *atomic sentences*. Then, a *language* $\mathcal{L}_{\text{gCL}}(\mathcal{I})$ for graded conditional logic is formed from $\mathcal{I}$ as the set of sentences closed under the usual propositional operators such as $\top, \bot, \neg, \wedge, \vee, \rightarrow$, and $\leftrightarrow$ as well as $\square\!\!\!\rightarrow_k$ and $\Diamond\!\!\!\rightarrow_k$ (*graded conditionals*) for $0 < k \leq 1$ in the usual way.

1. If $x \in \mathcal{I}$ then $x \in \mathcal{L}_{\text{gCL}}(\mathcal{I})$.
2. $\top, \bot \in \mathcal{L}_{\text{gCL}}(\mathcal{I})$.
3. If $p \in \mathcal{L}_{\text{gCL}}(\mathcal{I})$ then $\neg p \in \mathcal{L}_{\text{gCL}}(\mathcal{I})$.
4. If $p, q \in \mathcal{L}_{\text{gCL}}(\mathcal{I})$ then $p \wedge q, p \vee q, p \rightarrow q, p \leftrightarrow q \in \mathcal{L}_{\text{gCL}}(\mathcal{I})$,
5. If $[p, q \in \mathcal{L}_{\text{gCL}}(\mathcal{I})$ and $0 < k \leq 1]$ then $p\square\!\!\!\rightarrow_k q, p\Diamond\!\!\!\rightarrow_k q \in \mathcal{L}_{\text{gCL}}(\mathcal{I})$.

A measure-based conditional model is defined as a family of minimal conditional model (cf. Chellas [7]):

Definition 9.7 Given a database $\mathcal{D} = \langle T, V \rangle$ on $\mathcal{I}$ and a fuzzy measure m, a *measure-based conditional model* $\mathcal{M}_{\text{g}\mathcal{D}}^m$ for $\mathcal{D}$ is a structure

$$\langle W, \{g_k\}_{0<k\leq1}, v \rangle,$$

where (1) $W = T$, (2) for any world (transaction) t in W and any set of itemsets X in $2^{\mathcal{I}}$, g_k is defined by a fuzzy measure m as

$$g_k(t, X) \stackrel{\text{def}}{=} \{Y \subseteq 2^W \mid m(Y, X) \geq k\},$$

and (3) for any item x in $\mathcal{I}$, $v(x, t) = 1$ iff $x \in V(t)$. ∎

The model $\mathcal{M}_{\text{g}\mathcal{D}}^m$ is called *finite* because so is W. Further, in this paper, we call the model $\mathcal{M}_{\text{g}\mathcal{D}}^m$ *uniform* since functions $\{g_k\}$ in the model does not depend on any world in $\mathcal{M}_{\text{g}\mathcal{D}}^m$.

The truth condition for $\square\!\!\!\rightarrow_k$ in a grade conditional model is given by

$$\mathcal{M}_{\text{g}\mathcal{D}}^m, t \models p\square\!\!\!\rightarrow_k q \text{ iff } \|q\|^{\mathcal{M}_{\text{g}\mathcal{D}}^m} \in g_k(t, \|p\|^{\mathcal{M}_{\text{g}\mathcal{D}}^m}),$$

where

$$\|p\|^{\mathcal{M}_{\text{g}\mathcal{D}}^m} \stackrel{\text{def}}{=} \{t \in W(= T) \mid \mathcal{M}_{\text{g}\mathcal{D}}^m, t \models p\}.$$

The basic idea of this definition is the same as in fuzzy-measure-based semantics for graded modal logic defined in [12–14].

9.4.3 Association Rules and Graded Conditionals

For example, the usual degree of confidence [1] is nothing but the well-known conditional probability, so we define function g_k by conditional probability.

Definition 9.8   For a given database $\mathcal{D} = \langle T, V \rangle$ on $\mathcal{I}$ and a conditional probability

$$\Pr(B|A) = \frac{|A \cap B|}{|A|},$$

its corresponding *probability-based graded conditional model* $\mathcal{M}_{g\mathcal{D}}^{\mathrm{Pr}}$ is defined as a structure

$$\langle W, \{g_k\}_{0 < k \leq 1}, v \rangle,$$

where

$$g_k(w, X) \overset{\text{def}}{=} \{Y \subseteq 2^W \mid \Pr(t(Y) \mid t(X)) \geq k\},$$

where

$$t(X) \overset{\text{def}}{=} \{w \in W \mid X \subseteq w\}. \ \blacksquare$$

The truth condition of graded conditional is given by

$$\mathcal{M}_{g\mathcal{D}}^{\mathrm{Pr}}, t \models p\Box\!\!\!\rightarrow_k q \text{ iff } \Pr(\|q\|^{\mathcal{M}_{g\mathcal{D}}^{\mathrm{Pr}}} \mid \|p\|^{\mathcal{M}_{g\mathcal{D}}^{\mathrm{Pr}}}) \geq k.$$

Then we can have the following theorem:

Theorem 9.1 *Given a database $\mathcal{D}$ on $\mathcal{I}$ and its corresponding probability-based graded conditional model $\mathcal{M}_{g\mathcal{D}}$, for an association rule $X \Longrightarrow Y$, we have*

$$c(X \Longrightarrow Y) \geq k \text{ iff } \mathcal{M}_{g\mathcal{D}}^{\mathrm{Pr}} \models p_X\Box\!\!\!\rightarrow_k p_Y. \ \blacksquare$$

9.4.4 Paraconsistency and Paracompleteness in Association Rules

We formulated association rules as graded conditionals based on probability. Define the following set of rules with confidence k:

$$R_k \overset{\text{def}}{=} \{p\Box\!\!\!\rightarrow_k q \in \mathcal{L}_{gCD} \mid \vdash_{gCL} p\Box\!\!\!\rightarrow_k q\}.$$

A graded conditional $p\Box\!\!\!\rightarrow_k q$ is also regarded as a relative necessary sentences $[p]_k q$ and the properties of relative modal operator $[\cdot]_k$ are examined in Murai et al. [12, 13], [14] in the following correspondence:

Confidence k	Systems
$0 < k \leq \frac{1}{2}$	EMD_CNP
$\frac{1}{2} < k < 1$	$EMDNP$
$k = 1$	KD

The former two systems are not regular, so R_k may be paraconsistent. The last one is normal so regular.

For $0 < k \leq \frac{1}{2}$, R_k is complete but for some p and q, the both rules $p \Box \mapsto_k q$ and $p \Box \mapsto_k \neg q$ may be generated. This should be avoided.

For $\frac{1}{2} < k < 1$, R_k is consistent but may be paracomplete.

9.5 Dempster-Shafer-Theory-Based Confidence

9.5.1 D-S Theory and Confidence

The standard confidence [1] described in the previous section is based on the idea that co-occurrences of items in one transaction are evidence for association between the items. Since the definition of confidence is nothing but a conditional probability, even weights are a priori assigned to each transaction that contains the items in question at the same time. All of such transactions, however, do not necessarily give us such evidence because some co-occurrences might be contingent. Thus we need a framework that can differentiate proper evidence from contingent one and we introduce Dempster-Shafer theory of evidence [9, 19] to describe such a more flexible framework to compute confidence. There are a variety of ways of formalizing D-S theory and, in this paper, we adopt multivalued-mapping-based approach, which was originally used by Dempster [9]. In the approach, we need two frames, one of which has a probability defined, and a multivalued mapping between the two frames. Given a database $\mathcal{D} = \langle T, V \rangle$ on $\mathcal{I}$ and an association rule $r = (X \implies Y)$ in $\mathcal{D}$, one of frames is the set T of transactions. Another one is defined by

$$R = \{r, \bar{r}\},$$

where $\bar{r}$ denotes the negation of r. The remaining tasks are (1) to define a probability distribution Pr on T: $\text{Pr} : T \to [0, 1]$, and (2) to define a multivalued mapping $\Gamma :$ $T \to 2^R$. Given Pr and Γ, we can define the well-known two kinds of functions in Dempster-Shafer theory: for $X \subseteq 2^R$,

$$\text{Bel}(X) \overset{\text{def}}{=} \text{Pr}(\{t \in T \mid \Gamma(t) \subseteq X\}),$$

$$\text{Pl}(X) \overset{\text{def}}{=} \text{Pr}(\{t \in T \mid \Gamma(t) \cap X \neq \emptyset\})$$

which are called *belief* and *plausibility* functions, respectively. Now we have the following double-indexed confidence:

$$c(r) = \langle \text{Bel}(r), \text{Pl}(r) \rangle.$$

9.5.2 Multi-graded Conditional Models for Databases

Given a finite set $\mathcal{I}$ of items as *atomic sentences*, a *language* $\mathcal{L}_{\text{mgCL}}(\mathcal{I})$ for graded conditional logic is formed from $\mathcal{I}$ as the set of sentences closed under the usual propositional operators as well as $\Box\!\!\to_k$ and $\Diamond\!\!\to_k$ (*graded conditionals*) for $0 < k \leq 1$ in the usual way. Note that, in particular,

$$(p, q \in \mathcal{L}_{\text{mgCL}}(\mathcal{I}) \text{ and } 0 < k \leq 1) \Rightarrow p\Box\!\!\to_k q, p\Diamond\!\!\to_k q \in \mathcal{L}_{\text{mgCL}}(\mathcal{I}).$$

Definition 9.9 Given a database $\mathcal{D}$ on $\mathcal{I}$, a *multi-graded conditional model* $\mathcal{M}_{\text{mg}\mathcal{D}}$ for $\mathcal{D}$ is a structure

$$\langle W, \{\{\underline{g}_k, \overline{g}_k\}\}_{0<k\leq 1}, v \rangle,$$

where (1) $W = T$, (2) for any world (transaction) t in W and any set of itemsets $\mathcal{X}$ in $2^{\mathcal{I}}$, g_k is defined by belief and plausibility functions:

$$\underline{g}_k(t, \mathcal{X}) \stackrel{\text{def}}{=} \{\mathcal{Y} \subseteq 2^W \mid \text{Bel}(\mathcal{Y}, \mathcal{X}) \geq k\},$$
$$\overline{g}_k(t, \mathcal{X}) \stackrel{\text{def}}{=} \{\mathcal{Y} \subseteq 2^W \mid \text{Pl}(\mathcal{Y}, \mathcal{X}) \geq k\},$$

and (3) for any item x in $\mathcal{I}$, $v(x, t) = 1$ iff $x \in V(t)$ $\blacksquare$

The truth conditions for $\Box\!\!\to_k$ and $\Diamond\!\!\to_k$ are given by

$$\mathcal{M}_{\text{mg}\mathcal{D}}, w \models p\Box\!\!\to_k q \text{ iff } \|q\|^{\mathcal{M}_{\text{mg}\mathcal{D}}} \in \underline{g}_k(t, \|p\|^{\mathcal{M}_{\text{mg}\mathcal{D}}})$$
$$\mathcal{M}_{\text{mg}\mathcal{D}}, w \models p\Diamond\!\!\to_k q \text{ iff } \|q\|^{\mathcal{M}_{\text{mg}\mathcal{D}}} \in \overline{g}_k(t, \|p\|^{\mathcal{M}_{\text{mg}\mathcal{D}}}),$$

respectively. Its basic idea is also the same as in fuzzy-measure-based semantics for graded modal logic defined in [12–14]. Several soundness results based on belief- and plausibility-function-based semantics (cf. [12–14]) are shown in Table 9.3.

9.5.3 Two Typical Cases

First we define a probability distribution on T by

$$\Pr(t) \stackrel{\text{def}}{=} \begin{cases} \frac{1}{a}, & \text{if } t \in \|p_X\|^{\mathcal{M}_{\text{mg}\mathcal{D}}}, \\ 0, & \text{otherwise,} \end{cases}$$

where $a = |\|p_X\|^{\mathcal{M}_{\text{mg}\mathcal{D}}}|$. This means that each world (transaction) t in $\|p_X\|^{\mathcal{M}_{\text{mg}\mathcal{D}}}$ is given an even mass (weight) $\frac{1}{a}$. To generalize the distribution is of course another interesting task.

Table 9.3 Soundness results of graded conditionals by belief and plausibility functions

Belief function			Rules and axiom schemata	Plausibility function		
$0 < k \le \frac{1}{2}$	$\frac{1}{2} < k < 1$	$k = 1$		$0 < k \le \frac{1}{2}$	$\frac{1}{2} < k < 1$	$k = 1$
○	○	○	RCEA	○	○	○
○	○	○	RCEC	○	○	○
○	○	○	RCM	○	○	○
○	○	○	RCR			○
○	○	○	RCN	○	○	○
		○	RCK			
○	○	○	CM	○	○	○
○	○	○	CC			○
○	○	○	CR			○
○	○	○	CN	○	○	○
○	○	○	CP	○	○	○
		○	CK			
	○	○	CD			
			CD_C	○		

Next we shall see two typical cases of definition of Γ. First we describe strongest cases. When we define a mapping Γ by

$$\Gamma(t) \stackrel{\text{def}}{=} \begin{cases} \{r\}, & \text{if } t \in \|p_X\|^{\mathcal{M}_{\mathrm{mg}\mathcal{D}}}, \\ \{\bar{r}\}, & \text{otherwise.} \end{cases}$$

This means that the transactions in $\|p_X \wedge p_Y\|^{\mathcal{M}_{\mathrm{mg}\mathcal{D}}}$ contribute as evidence to r, while the transactions in $\|p_X \wedge \neg p_Y\|^{\mathcal{M}_{\mathrm{mg}\mathcal{D}}}$ contribute as evidence to $\bar{r}$. This is the strongest interpretation of co-occurrences. Then, we can compute $\mathrm{Bel}(r) = \frac{1}{a} \times b$ and $\mathrm{Pl}(r) = \frac{1}{a} \times b$, where $b = |\|p_X \wedge p_Y\|^{\mathcal{M}_{\mathrm{mg}\mathcal{D}}}|$. Thus the induced belief and plausibility functions collapse to the same probability measure Pr: $\mathrm{Bel}(r) = \mathrm{Pl}(r) = \mathrm{Pr}(r) = \frac{b}{a}$, and thus

$$c(r) = \langle \frac{b}{a}, \frac{b}{a} \rangle.$$

Hence this case represents the usual confidence. According to this idea, in our movie database, we can define Pr and Γ in the way in Fig. 9.1. That is, any movie in $\|AH \wedge HM\|^{\mathcal{M}_{\mathrm{mg}\mathcal{D}}}$ contributes as evidence to that the rule holds (r), while all movie in $\|AH \wedge \neg HM\|^{\mathcal{M}_{\mathrm{mg}\mathcal{D}}}$ contributes as evidence to that the rule does not hold ($\bar{r}$). Thus we have

$$c(\{AH\} \Longrightarrow \{HM\}) = \langle 0.5, 0.5 \rangle.$$

No.	Transaction (movie)	AH	HM	Pr
1	Secret people	1		$\frac{1}{8}$
2	Monte Carlo baby	1		$\frac{1}{8}$
3	Roman holiday	1		$\frac{1}{8}$
4	My fair lady	1		$\frac{1}{8}$
5	Breakfast at Tiffany's	1	1	$\frac{1}{8}$
6	Charade	1	1	$\frac{1}{8}$
7	Two for the road	1	1	$\frac{1}{8}$
8	Wait until dark	1	1	$\frac{1}{8}$
9	Days of wine and rose		1	0
10	The great race		1	0
11	The pink panther		1	0
12	Sunflower		1	0
13	Some like it hot			0
14	12 Angry men			0
15	The apartment			0
......				
100	Les aventuriers			0

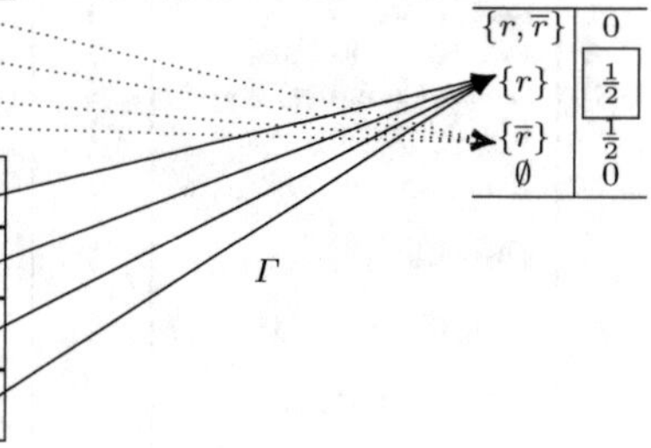

$\{r,\overline{r}\}$	0
$\{r\}$	$\frac{1}{2}$
$\{\overline{r}\}$	$\frac{1}{2}$
$\emptyset$	0

Γ

Fig. 9.1 An example of the strongest cases

Next we describe weakest cases. In general, co-occurrences do not necessarily mean actual association. The weakest interpretation of co-occurrences is to consider transactions totally unknown as described as follows: When we define a mapping Γ by

$$\Gamma(t) \stackrel{\text{def}}{=} \begin{cases} \{r, \overline{r}\}, & \text{if } t \in \|p_X\|^{\mathcal{M}_{\text{mg}\mathcal{D}}}, \\ \{\overline{r}\}, & \text{otherwise.} \end{cases}$$

This means that the transactions in $\|p_X \wedge p_Y\|^{\mathcal{M}_{\text{g}\mathcal{D}}}$ contribute as evidence to $R = \{r, \overline{r}\}$, while the transactions in $\|p_X \wedge \neg p_Y\|^{\mathcal{M}_{\text{g}\mathcal{D}}}$ contribute as evidence to $\overline{r}$. Then, we can compute $\text{Bel}(r) = 0$ and $\text{Pl}(r) = \frac{1}{a} \times b$, and thus

$$c(r) = \langle 0, \frac{b}{a} \rangle.$$

According to this idea, in our movie database, we can define Pr and Γ in the way in Fig. 9.2. That is, all movie in $\|AH \wedge \neg HM\|^{\mathcal{M}_{\text{mg}\mathcal{D}}}$ contributes as evidence to that the rule does not hold ($\overline{r}$), while we cannot expect whether each movie in $\|AH \wedge HM\|^{\mathcal{M}_{\text{mg}\mathcal{D}}}$ contributes or not as evidence to that the rule holds (r). Thus we have

$$c(\{AH\} \Longrightarrow \{HM\}) = \langle 0, 0.5 \rangle.$$

In the case, the induced belief and plausibility functions, denoted respectively $Bel_{bpa'}$ and $Pl_{bpa'}$, become *necessity* and *possibility* measures in the sense of Dubois and Prade [10]. We have several soundness results based on necessity- and possibility-measure-based semantics (cf. [12–14]) shown in Table 9.4.

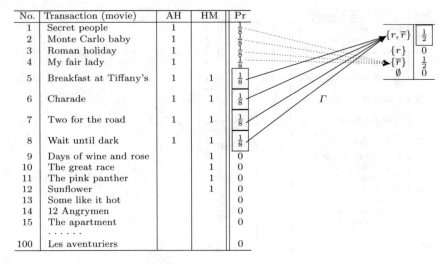

No.	Transaction (movie)	AH	HM	Pr
1	Secret people	1		$\frac{1}{8}$
2	Monte Carlo baby	1		$\frac{1}{8}$
3	Roman holiday	1		$\frac{1}{8}$
4	My fair lady	1		$\frac{1}{8}$
5	Breakfast at Tiffany's	1	1	$\frac{1}{8}$
6	Charade	1	1	$\frac{1}{8}$
7	Two for the road	1	1	$\frac{1}{8}$
8	Wait until dark	1	1	$\frac{1}{8}$
9	Days of wine and rose		1	0
10	The great race		1	0
11	The pink panther		1	0
12	Sunflower		1	0
13	Some like it hot			0
14	12 Angrymen			0
15	The apartment			0
......				
100	Les aventuriers			0

Fig. 9.2 An example of the weakest cases

Table 9.4 Soundness results of graded conditionals by necessity and possibility measures

Necessity measure $0 < k \leq 1$	Rules and axiom schemata	Possibility measure $0 < k \leq 1$
O	RCEA	O
O	RCEC	O
O	RCM	O
O	RCR	
O	RCN	O
O	RCK	
O	CM	O
O	CC	
	CF	O
O	CR	O
O	CN	O
O	CP	O
O	CK	
O	CD	
	CD_C	O

9.5.4 General Cases

In the previous two typical cases, one of which coincides to the usual confidence, any transaction in $\|AH \wedge HM\|^{\mathcal{M}_{mg}\mathcal{D}}$ (or in $\|AH \wedge \neg HM\|^{\mathcal{M}_{mg}\mathcal{D}}$) has the same weight as evidence. It would be, however, possible that some of $\|AH \wedge HM\|^{\mathcal{M}_{mg}\mathcal{D}}$

No.	Transaction (movie)	AH	HM	Pr
1	Secret people	1		$\frac{1}{8}$
2	Monte Carlo baby	1		$\frac{1}{8}$
3	Roman holiday	1		$\frac{1}{8}$
4	My fair lady	1		$\frac{1}{8}$
5	Breakfast at Tiffany's	1	1	$\frac{1}{8}$
6	Charade	1	1	$\frac{1}{8}$
7	Two for the road	1	1	$\frac{1}{8}$
8	Wait until dark	1	1	$\frac{1}{8}$
9	Days of wine and rose		1	0
10	The great race		1	0
11	The pink panther		1	0
12	Sunflower		1	0
13	Some like it hot			0
14	12 Angry men			0
15	The apartment			0
	. . . . . .			
100	Les aventuriers			0

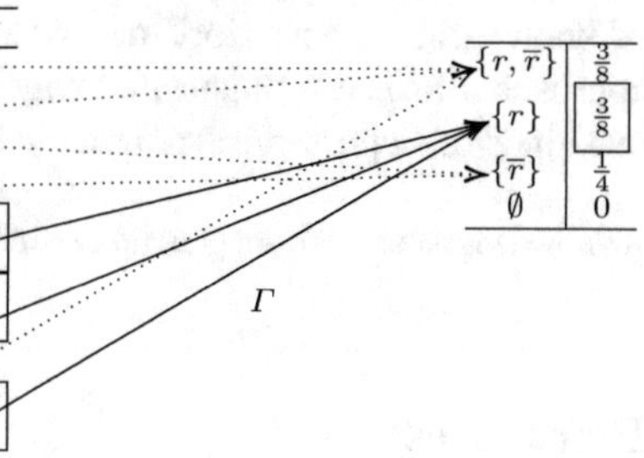

Diagram (right side, Γ):
$\{r, \bar{r}\}$ $\frac{3}{8}$; $\{r\}$ $\frac{3}{8}$; $\{\bar{r}\}$ $\frac{1}{4}$; $\emptyset$ 0

Fig. 9.3 An example of general cases

(or $\|AH \wedge \neg HM\|^{\mathcal{M}_{\mathrm{mg}\mathcal{D}}}$) does work as positive evidence to r (or $\bar{r}$) but other part does not.

Thus we have a tool that allows us to introduce various kinds of 'a posteriori' pragmatic knowledge into the logical setting of association rules. As an example, we assume that (1) the music of the first and second movies was not composed by Mancini, but the fact does not affect the validity of $\bar{r}$ because they are not very important ones, and (2) the music of the seventh movie was composed by Mancini, but the fact does not affect the validity of r. Then we can define Γ in the way in Fig. 9.3. Thus we have

$$c(\{AH\} \Longrightarrow \{HM\}) = \langle 0.375, 0.75 \rangle.$$

In general, users have such kind of knowledge 'a posteriori.' Thus the D-S based approach allows us to introduce various kinds of 'a posteriori' pragmatic knowledge into association rules.

9.6 Concluding Remarks

In this paper, we examined paraconsistency and paracompleteness that appear in association rules in a framework of probability-based models for conditional logics. For lower values of confidence (less than or equal to $\frac{1}{2}$), both $p\square\mapsto_k q$ and $p\square\mapsto_k \neg q$ may be generated so we must be careful to use such lower confidence.

Further we extended the above discussion into the case of Dempster-Shafer theory of evidence to double-indexed confidences. Users have, in general, such kind of knowledge 'a posteriori' describe in the previous section. Thus the D-S based approach allows a sophisticated way of calculating confidence by introducing various kinds of 'a posteriori' pragmatic knowledge into association rules.

Acknowledgments We are grateful to a referee for useful comments.

References

1. Agrawal, R., Imielinski, T., Swami, A.: Mining association rules between sets of items in large databases. In: Proceedings of the ACM SIGMOD Conference on Management of Data, pp. 207–216 (1993)
2. Agrawal, R., Mannila, H., Srikant, R., Toivonen, H., Verkamo, A.I.: Fast discovery of association rules. In: Fayyad, U.M., Platetsky-Shapiro, G., Smyth, P., Uthurusamy, R. (eds.) Advances in Knowledge Discovery and Data Mining, pp. 307–328. AAAI Press/The MIT Press (1996)
3. Aggarwal, C.C., Philip, S.Y.: Online generation of association rules. In: Proceedings of the International Conference on Data Engineering, pp. 402–411 (1998)
4. Akama, S., Abe, J.M.: Many-valued and annotated modal logics. In: Proceedings of 28th ISMVL, pp. 114–119 (1998)
5. Akama, S., Abe, J.M.: Fuzzy annotated logics. In: Proceedings of IPMU 2000, pp. 504–509 (2000)
6. Akama, S., Abe, J.M.: Paraconsistent logics viewed as a foundation of data warehouses. Advances in Logic, Artificial Intelligence and Robotics, pp. 96–103. IOS Press (2002)
7. Chellas, B.F.: Modal Logic: An Introduction. Cambridge University Press, Cambridge (1980)
8. da Costa, N.C.A., Abe, J.M., Subrahmanian, V.S.: Remarks on annotated logic. Zeitschr. f. Math. Logik und Grundlagen d. Math. **37**, 561–570 (1991)
9. Dempster, A.P.: Upper and lower probabilities induced by a multivalued mapping. Ann. Math. Stat. **38**, 325–339 (1967)
10. Dubois, D., Prade, H.: Possibility Theory: An Approach to Computerized Processing of Uncertainty. Springer (1988)
11. Lewis, D.: Counterfactuals. Blackwell, Oxford (1973)
12. Murai, T., Miyakoshi, M., Shimbo, M.: Measure-based semantics for modal logic. In: Lowen, R., Roubens, M. (eds.) Fuzzy Logic: State of the Art, pp. 395–405. Kluwer, Dordrecht (1993)
13. Murai, T., Miyakoshi, M., Shimbo, M.: Soundness and completeness theorems between the Dempster-Shafer theory and logic of belief. In: Proceedings of the 3rd FUZZ-IEEE (WCCI), pp. 855–858 (1994)
14. Murai, T., Miyakoshi, M., Shimbo, M.: A logical foundation of graded modal operators defined by fuzzy measures. In: Proceedings of the 4th FUZZ-IEEE/2nd IFES, pp. 151–156 (1995)
15. Murai, T., Sato, Y.: Association rules from a point of view of modal logic and rough sets. In: Proceedings of the 4th AFSS, pp. 427–432 (2000)
16. Murai, T., Nakata, M., Sato, Y.: A note on conditional logic and association rules. In: Terano, T., et al. (eds.) New Frontiers in Artificial Intelligence, LNAI, vol. 2253, pp. 390–394. Springer (2001)
17. Murai, T., Nakata, M., Sato, Y.: Association rules as relative modal sentences based on conditional probability. Commun. Inst. Inf. Comput. Mach. **5**(2), 73–76 (2002)
18. Murai, T., Sato, Y., Kudo, Y.: Paraconsistency and neighborhood models in modal logic. In: Proceedings of the 7th World Multiconference on Systemics, Cybernetics and Informatics, vol. XII, pp. 220–223 (2003)
19. Shafer, G.: A Mathematical Theory of Evidence. Princeton University Press (1976)

Chapter 10
A Beautiful Theorem

Francisco Antonio Doria and Carlos A. Cosenza

Dedicated to Jair Minoro Abe for his 60th birthday

A thing of beauty is a joy for ever:
Its loveliness increases; it will never
Pass into nothingness

John Keats, Endymion

Abstract We first present Maymin's Theorem on the existence of efficient markets; it is a result that connects mathematical economics and computer science. We then introduce O'Donnell's algorithm for the solution of NP-complete problems and the concept of almost efficient markets; we state the main result, which is: given a metamathematical condition, there will be almost efficient markets. We then briefly discuss whether changing the underlying logical framework we would be able to change the preceding results.

Keywords Maymin's theorem · Efficient markets · O'Donnell's algorithm · Almost efficient markets

10.1 Prologue

Beauty in mathematics has many sources. One of them, the discovery of links between areas that seemed at first so far away. That's the case of Maymin's theorem on efficient markets: it is linked as in the gesture of a magician to complexity theory in computer

Partially supported by CNPq, Philosophy Section; the first author is a member of the Brazilian Academy of Philosophy.

F.A. Doria (✉) · C.A. Cosenza
Advanced Studies Research Group, HCTE, Fuzzy Sets Laboratory,
Mathematical Economics Group, Production Engineering Program,
COPPE, UFRJ, P.O. Box 68507, Rio Rj 21945–972, Brazil
e-mail: fadoria63@gmail.com

© Springer International Publishing Switzerland 2016
S. Akama (ed.), *Towards Paraconsistent Engineering*, Intelligent Systems
Reference Library 110, DOI 10.1007/978-3-319-40418-9_10

science. An apt paraphrasis for Maymin's result would be—there are weakly efficient markets if and only if a given major question in computer science is trivialized.

The major question is, of course, the *P* versus *NP* question.

The proof is simple, once we see that markets (in Maymin's sense) are naturally coded as Boolean formulae.

The present paper first sketches that construction. We then fill in the required details and prove Maymin's theorem. After that we construct O'Donnell's algorithm, show that it is near–polynomial given reasonable conditions, and then with the help of that algorithm we define "almost Maymin" efficient markets. We then prove the existence of almost Maymin efficient markets, again given a reasonable metamathematical hypothesis.

Main sources for these results are [3, 4]. And let's repeat here our motto: *A thing of beauty is a joy forever.* So is Maymin's theorem.

10.2   Theme

Think of this paper as a set of variations over a theme found elsewhere. We will pick up our main theme from two sources, Maymin's original paper [4] and a summary of it made in a recent paper by the author (with NCA da Costa) [2, 3, 5].

A Brief Scenario

We start here from this recent intriguing result by Maymin [4]. We use a modified, restricted version of Maymin's construction (but there is no loss of generality in our construction.) The concepts we require are that of a *Maymin market*, soon to be clarified.

Roughly, a Maymin market is a market coded by a Boolean expression, as we will see. Despite this very precise identification, the object we consider is quite general. Basically we are going to make some move in the market. Our move now is determined by a series of k previous moves. More precisely:

Definition 10.2.1 • A k-run policy σ_k, k a positive integer, is a series of plays (b for buy and s for sell) of length k. There are clearly 2^k possible k-run policies.
• A map v from all possible k-run policies into $\{0, 1\}$ is a valuation; we have a "gain" iff $v(\sigma_k) = 1$; a "loss" otherwise.
• A policy is *successful* if it provides some gain (adequately defined); in that case we put $v(\sigma_k) = 1$. Otherwise $v(\sigma_k) = 0$.   □

There is a natural map between these objects and k-variable Boolean expressions (see below), if we take that $v(\sigma_k) = 1$ means that σ_k is satisfiable, and 0 otherwise. We say that a *market configuration* (k-steps market configuration, or simply k-market configuration) is coded by a Boolean expression in disjunctive normal form (dnf).

That map between k-market configurations and k-variable Boolean expressions in dnf can be made 1–1.

The financial game for our simplified market is simple: we wish to discover the fastest way to algorithmically obtain a successful k-market configuration, given a particular market (i.e., a given k-variable Boolean expression).

Finally the k-market configurations are Maymin–efficient (see below) if v can be implemented by a poly algorithm.

Clearly there is a general polynomial procedure to do it if and only if $P = NP$. From what we know about the P versus NP question,[1] in particular cases we can of course find polynomial procedures, but it is unknown whether there are general procedures that are polynomial.

Stretto

Maymin restricts his analysis to the so-called "weakly efficient" markets. Since he adds the condition that there is a time-polynomial algorithmic procedure to spread the data about the market, we name *Maymin–efficient markets* those markets, where (we stress) $v(\sigma_k)$ is computed by a time-polynomial Turing machine (or poly-machine).

So the existence of general poly procedures characterizes the market as *Maymin efficient*. We can therefore state Maymin's theorem:

Proposition 10.2.1 *Markets are (Maymin) efficient if and only if $P = NP$.*   □

Now we put: markets are *almost Maymin efficient* if and only if there is an O'Donnell algorithm to determine its successful policies [3]. Then:

Proposition 10.2.2 *If $P < NP$ isn't proved by primitive recursive arithmetic then there are almost Maymin efficient markets.*   □

We are now going to expand these brief remarks into a detailed proof of Maymin's theorem, and then add to it some spice of our own.

10.3   Theme and Variations

The main motive is very simple: we are going to code Maymin–efficient markets as Boolean expressions. This is the main trick. But how do we proceed?

We first require a classical result by Emil Post [6]. The 2^k binary sequences naturally code integers from 0 to $2^k - 1$; more precisely, from:

$$000\ldots00, k \text{ digits},$$

to:

$$111\ldots11, k \text{ digits}.$$

Fix one such coding; a k-digit binary sequence is seen as a sequence of truth values for a Boolean expression Ξ_k. After we test the Boolean expression with one

---

[1] Actually we know very little.

specific line of truth values, we see whether that particular line satisfies or doesn't satisfy Ξ_k.

Proposition 10.3.1 *Let ξ_k be a binary sequence of length 2^k. Then there is a Boolean expression Ξ_k on k Boolean variables so that ξ_k is its truth table.*

(We take 1 as "true" and 0 as "false."). The idea of the proof goes as follows. Notice that the Boolean expression:

$$\neg p_1 \wedge p_2 \wedge p_3 \wedge \neg p_4 \wedge \neg p_5$$

is satisfied by the binary 5-digit line:

$$01100$$

(When there is a $\neg$ in the conjunction put 0 in the line of truth-values; if not put 1.)

The line 01100 satisfies the Boolean conjunction above, and no other 5-digit line will satisfy it, that is, it has a truth table where a single line of truth values satisfies it—the truth table has a single 1 and is zero for all remaining truth value lines.

In order to obtain a truth table with just two 1's, we construct the conjuncts as above that have the desired lines and then get the expression which is the disjunction of those conjuncts. That is the idea in the proof of Post's theorem.

Trivially every k-variable Boolean expression gives rise to a 2^k-length truth table which we can code as a binary sequence of, again, size 2^k bits. The converse result is given by Post's theorem.

Proof of Post's theorem, a sketch: Consider the k-variable Boolean expression:

$$\zeta = \alpha_1 p_1 \wedge \alpha_2 p_2 \wedge \cdots \wedge \alpha_k p_k,$$

where the α_i are either nothing or $\neg$. Pick up the line of truth values $\zeta' = \alpha_1 \alpha_2 \ldots \alpha_k$, where "nothing" stands for 1 and $\neg$ for 0. ζ' satisfies ζ, while no other line of truth values does. Our Boolean expression ζ is satisfied by ζ' and by no other k-digit line of truth values.

The disjunction $\zeta \vee \xi$ where ξ is a k-variable Boolean expression as ζ, is satisfied by (correspondingly) two lines of truth values, and no more. And so on.

The rigorous proof of Post's theorem is by finite induction. □

Now:

Definition 10.3.1 The Boolean expression in dnf ζ is identified to a Maymin k-market configuration. □

Then:

Proposition 10.3.2 *There are Maymin–efficient markets if and only if $P = NP$.*

Proof Such is the condition for the existence of a poly algorithmic map ν. □

10.4   The O'Donnell Algorithm

We are now going to describe O'Donnell's algorithm [2, 3, 5]; the O'Donnell algorithm is a quasi-polynomial algorithm for SAT.[2] We require the so-called BGS set of poly machines and f_c, which is the (now recursive) counterexample function to $[P = NP]$ (See [1, 3] for details.)

Remark 10.4.1 A BGS machine is a Turing machine $M_n(x)$ coupled to a clock that stops the machine when it has operated for $|x|^p + p$ steps, where x is the binary input to the machine and $|x|$ is its length in bits; p is an integer ≥ 1. Of course the coupled system is a Turing machine. All machines in the BGS set are poly machines, and given any poly machine, there will be a corresponding machine in BGS with the same output as the original poly machine.   □

Remark 10.4.2 f_c is the recursive counterexample function to $P = NP$. To get it:

- Enumerate all BGS machines in the natural order (one can do it, as the BGS set is recursive).
- For BGS machine P_n, $f_c(n)$ equals the first instance of SAT which is input to P_n and fails to output a satisfying line for that instance of SAT.   □

O'Donnell's algorithm is very simple: we list in the natural ordering all BGS machines. Given a particular instance $x \in$ SAT, we input it to $P_1, P_2, \ldots$ up to the moment when the output is a satisfying line of truth values. When we compute the time bound to that procedure, we see that it is near polynomial, that is, the whole operation is bounded by a very slow-growing exponential.

Now some requirements:

- We use the (fixed) enumeration of finite binary sequences

$$0, 1, 00, 01, 10, 11, 000, 001, 010, 011, \ldots.$$

If *FB* denotes the set of all such finite binary sequences, form the standard coding $FB \mapsto \omega$ which is monotonic on the length of the binary sequences.
- We use a binary coding for the Turing machines which is also monotonic on the length of their tables, linearly arranged, that is, a 3-line table s_1, s_2, s_3, becomes the line $s_1 - s_2 - s_3$.
 We call such monotonic codings *standard codings*.
- We consider the set of all Boolean expressions in cnf,[3] including those that are unsatisfiable, or totally false. We give it the usual coding which is 1–1 and onto ω.
- Consider the poly Turing machine $V(x, s)$, where $V(x, s) = 1$ if and only if the binary line of truth values s satisfies the Boolean cnf expression x, and $V(x, s) = 0$ if and only if s doesn't satisfy x.

---

[2] Actually we deal with a slightly larger class of Boolean expressions.
[3] Conjunctive normal form.

- Consider the enumeration of the BGS [1] machines, $P_0, P_1, P_2, \ldots$.[4]

  We start from x, a Boolean expression in cnf binarily coded:

- Consider x, the binary code for a Boolean expression in cnf form.
- Input x to $P_0, P_1, P_2, \ldots$ up to the first P_j so that $P_j(x) = s_j$ and s_j satisfies x (that is, for the verifying machine $V(x, s_j) = 1$).
- Notice that there is a bound $\leq j = f_c^{-1}(x)$.

  This requires some elaboration. Eventually a poly machine (in the BGS sequence) will produce a satisfying line for x as its output given x as input. The upper bound for the machine with that ability is given by the first BGS index so that the code for x is smaller than the value at that index of the counterexample function.

  That means: we arrive at a machine M_m which outputs a correct satisfying line up to x as an input, and then begins to output wrong solutions.

- Alternatively check for $V(x, 0)$, $V(x, 1)$, …up to—if it ever happens—some s so that $V(x, s) = 1$; or,
- Now, if f_c is fast-growing, then as the operation time of P_j is bounded by $|x|^k + k$, we have that $k \leq j$, and therefore it grows as $O(f_c^{-1}(x))$. This will turn out to be a very slowly growing function.

  Again this requires some elaboration. The BGS machines are coded by a pair $\langle m, k \rangle$, where m is a Turing machine Gödel index, and k is as above. So we will have that the index j by which we code the BGS machine among all Turing machines is greater than k, provided we use a monotonic coding.

  More precisely, it will have to be tested up to j, that is the operation time will be bounded by $f_c^{-1}(x)(|x|^{f_c^{-1}(x)} + f_c^{-1}(x))$.

  Again notice that the BGS index $j \geq k$, where k is the degree of the polynomial clock that bounds the poly machine.

10.5 Almost Maymin–Efficient Markets

We will now discuss the following:

Proposition 10.5.1 *If $P < NP$ isn't proved by primitive recursive arithmetic then there are almost Maymin efficient markets.*   □

For a theory S with enough arithmetic—we leave it vague; we'll specify how much arithmetic in the example we'll soon discuss—and with a recursively enumerable set of theorems, for any provably total recursive function h there is a recursive, total, function g so that g dominates h.

Suppose now that we conjecture: the formal sentence $P < NP$ isn't proved by Primitive Recursive Arithmetic. Then the counterexample function f_c will be at least of the order of growth of Ackermann's function [3]. By the previous discussion about

---

[4]The BGS machine set is a set of time-polynomial Turing machines which includes algorithms that mimic all time-polynomial Turing machines. See above and check [1].

O'Donnell's algorithm, we see that the slow–growing exponential that bounds the operation time of the algorithm will be at least of the order of growth of the inverse function of Ackermann's function.

Given that condition, we can state:

Proposition 10.5.2 *If P < NP isn't proved by Primitive Recursive Arithmetic then there are almost Maymin–efficient markets.*   □

Comments

For details, [3]. We've briefly remarked that our proof, while restricted to a very particular situation, is in fact adequately general. That is the case: for a more general set of objects (policies etc.) has to imply ours, as our domain of objects would be a subset of the enlarged domain.

Also we require very little in our discussion—main tool is Post's theorem. As long as it holds, so does our proof. Does it hold for paraconsistent logics? That's an open question, which classes of paraconsistent logics would allow a proof of Maymin's beautiful theorem.

Acknowledgments This paper was supported in part by CNPq, Philosophy Section Grant no. 4339819902073398. It is part of the research efforts of the Advanced Studies Group, Production Engineering Program, at COPPE–UFRJ and of the Logic Group, HCTE–UFRJ. We thank Profs. R. Bartholo, S. Fuks (in memoriam), S. Jurkiewicz, R. Kubrusly, and F. Zamberlan for support.

References

1. Baker, T., Gill, J., Solovay, R.: Relativizations of the $P =?NP$ question. SIAM J. Comput. **4**, 431–442 (1975)
2. Ben–David, S., Halevi, S.: On the independence of P vs. NP. Technical Report # 699, Technion (1991)
3. da Costa, N.C.A., Doria, F.A.: On the O'Donnell algorithm for NP–complete problems. Rev. Behav. Econ. (2016)
4. Maymin, P.Z.: Markets are efficient if and only If $P = NP$. Algorithmic Finance, **1**(1), 1 (2011)
5. O'Donnell, M.: A programming language theorem which is independent of Peano arithmetic. In: Proceedings of 11th Annual ACM Symposium on the Theory of Computation, pp. 176–188 (1979)
6. Post, E.L.: Introduction to a general theory of elementary propositions. Am. J. Math. **43**, 163 (1921)

Chapter 11
Temporal Logic Modeling
of Biological Systems

Jean-Marc Alliot, Robert Demolombe, Martín Diéguez,
Luis Fariñas del Cerro, Gilles Favre, Jean-Charles Faye,
Naji Obeid and Olivier Sordet

Dedicated to Jair Minoro Abe for his 60th birthday

Abstract Metabolic networks, formed by a series of metabolic pathways, are made of intracellular and extracellular reactions that determine the biochemical properties of a cell, and by a set of interactions that guide and regulate the activity of these reactions. Cancer, for example, can sometimes appear in a cell as a result of some pathology in a metabolic pathway. Most of these pathways are formed by an intricate and complex network of chain reactions, and can be represented in a human readable form using graphs which describe the cell signaling pathways. In this paper, we define a logic, called Molecular Interaction Logic (MIL), able to represent these graphs and we present a method to automatically translate graphs into MIL formulas. Then we show how MIL formulas can be translated into linear time temporal logic, and then grounded into propositional classical logic. This enables us to solve complex queries on graphs using only propositional classical reasoning tools such as SAT solvers.

Keywords Metabolic networks · Molecular interaction logic (MIL) · Temporal reasoning

11.1 Introduction

Metabolic networks, formed by a series of metabolic pathways, are made of intracellular and extracellular reactions that determine the biochemical properties of a cell by consuming and producing proteins, and by a set of interactions that guide

J.-M. Alliot · R. Demolombe · M. Diéguez · L. Fariñas del Cerro (✉) ·
G. Favre · J.-C. Faye · N. Obeid · O. Sordet
INSERM/IRIT, University of Toulouse, Toulouse, France
e-mail: luis.farinas@irit.fr

© Springer International Publishing Switzerland 2016
S. Akama (ed.), *Towards Paraconsistent Engineering*, Intelligent Systems
Reference Library 110, DOI 10.1007/978-3-319-40418-9_11

205

and regulate the activity of these reactions. These reactions are at the center of a cell's existence, and are regulated by other proteins, which can either activate these reactions or inhibit them.

These pathways form an intricate and complex network of chain reactions, and can be represented in a human readable form using graphs which describe the cell signaling pathways.

These graphs can become extremely large, and although essential for knowledge capitalization and formalization, they are difficult to use:

- Reading is complex due to the very large number of elements, and reasoning is even more difficult.
- Using a graph to communicate goals is only partially suitable because the representation formalism requires expertise.
- Graphs often contain implicit knowledge, that is taken for granted by one expert, but is missed by another one.

Here, we show how classical propositional reasoning tools can be used to detect problems on these graphs, such as missing knowledge, and to answer complex queries.

The rest of this paper is organized as follows. Section 11.2 presents the important concepts and the problems to solve in layman's words with a simple example, Sect. 11.3 describes the concepts of production and regulation which are the basic operations present in a graph, Sect. 11.4 presents the Molecular Interaction Logic (MIL) capable of describing and reasoning about general pathways, Sect. 11.5 studies the relation between MIL and Linear Time Temporal Logic, Sect. 11.6 presents temporal reasoning and a method for grounding temporal theories into classical propositional formulas, when assuming bounded time, Sect. 11.7 explains what kind of queries on graphs can be answered using classical propositional reasoning tools such as SAT solvers, Sect. 11.8 describes the current state of the operational implementation of this tool, and at last Sect. 11.9 gives a summary and discusses future works.

11.2   A Simple Classical Example

We are first going to describe a simple graph, which represents the regulation of the *lac* operon.[1] A detailed presentation is available at [21].

The lac operon (lactose operon) is an operon required for the transport and metabolism of lactose in many bacteria. Although glucose is the preferred carbon source for most bacteria, the lac operon allows for the effective digestion of lactose when glucose is not available. The lac operon is a sequence of three genes (lacZ,

[1]The Nobel prize was awarded to Monod, Jacob and Lwoff in 1965 partly for the discovery of the lac operon by Monod and Jacob [16], which was the first genetic regulatory mechanism to be understood clearly, and is now a "standard" introductory example in molecular biology classes.

lacY and lacA) which encode 3 enzymes. Then, these enzyms carry the transformation of lactose into glucose. We will concentrate here on lacZ. LacZ encodes the β-galactosidase which cleaves lactose into glucose and galactose.

The lac operon uses a two-part control mechanism to ensure that the cell expends energy producing the enzymes encoded by the lac operon only when necessary. First, in the absence of lactose, the lac repressor halts production of the enzymes encoded by the lac operon. Second, in the presence of glucose, the catabolite activator protein (CAP), required for production of the enzymes, remains inactive.

Figure 11.1 describes this regulatory mechanism. The expression of lacZ gene is only possible when RNA polymerase (pink) can bind to a promotor site (marked P, black) upstream the gene. This binding is aided by the cyclic adenosine monophosphate (cAMP in blue) which binds before the promotor on the CAP site (dark blue).

The lacl gene (yellow) encodes the repressor protein Lacl (yellow) which binds to the promotor site of the RNA polymerase when lactose is not available, preventing the RNA polymerase to bind to the promoter and thus blocking the expression of the following genes (lacZ, lacY and lacA): this is a *negative regulation*, or *inhibition*, as it blocks the production of the proteins. When lactose is present, the repressor protein Lacl binds with lactose and is converted to allolactose, which is not able to

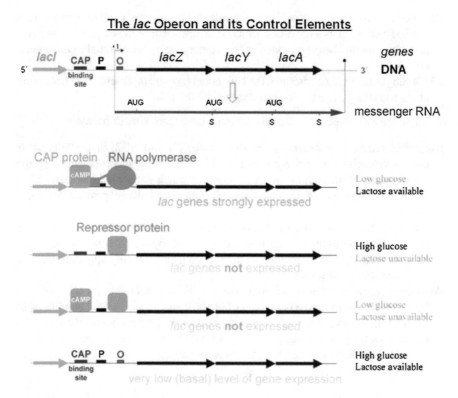

Fig. 11.1 Lac operon

bind to the promotor site, thus enabling RNA polymerase to bind to the promotor site and to start expressing the lacZ gene if cAMP is bound to CAP.

cAMP is on the opposite a *positive regulation*, or an *activation*, as its presence is necessary to express the lacZ gene. However, cAMP is itself regulated negatively by glucose: when glucose is present, the concentration of cAMP becomes low, and thus cAMP does not bind to the CAP site, blocking the expression of lacZ.

In this graph, we have three kinds of entities which have different initial settings and temporal dynamics:

- lacl, lacZ and cAMP are initial external conditions of the model and they do not evolve in time.
- galactosidase and the repressor protein can only be produced inside the graph, and are always absent at the start (time 0) of the modeling. Their value will then evolve in time according to the processes described by the graph.
- glucose and lactose also evolve in time (like galactosidase and the repressor protein) according to the processes described by the graph, but they are also initial conditions of the system, and can either be present or absent at time 0, like lacl, lacZ and cAMP.

So, an entity must be classified according to two main characteristics:

C1: It can evolve in time according to the cell reactions (appear and disappear), or it can be fixed, such as a condition which is independent of the cell reactions (temperature, protein always provided in large quantities by the external environment, etc…).

C2: It can be an initial condition of the cell model (present *or* absent at the beginning of the modeling), or can *only* be produced by the cell.

There are thus three kind of entities, which have three kind of behaviour:

Exogenous entities: an *exogenous* entity satisfies $C1$ and $\neg C2$; their status *never* change through time: they are set once and for all by the environment or by the experimenter at the start of the simulation; the graph never modifies their value, and if they are used in a reaction, the environment will always provide "enough" of them.

Pure endogenous entities: on the opposite, a *pure endogenous* entity satisfies $\neg C1$ and $C2$; their status evolves in time and is set *only* by the dynamic of the graph. They are absent at the beginning of the reaction, and can only appear if they are produced inside the graph.

Weak endogenous entities: *weak endogenous* entities satisfy $C2$ and $C1$; they can be present or absent at the beginning of the process (they are initial conditions of the model), however their value after the start of the process is entirely set by the dynamic of the graph. So they roughly behave like *pure endogenous* entities, but the initial condition can be set by the experimenter.

The status of a protein/condition is something which is set by the biologist, regarding his professional understanding of the biological process described by the graph.[2] However a rule of thumb is that exogenous entities are almost never produced inside the graph (they never appear at the right side of a production arrow), while endogenous entities always appear on the right side of a production arrow (but they can also appear on the left side of a production rule, especially weak endogenous entities).

These distinctions are fundamental, because the dynamics of these entities are different and they will have to be formalized differently.

11.3 Fundamental Operations

The mechanism described in the previous section is summarized in the simplified graph in Fig. 11.2. This example contains all the relationship operators that will be used in the rest of this document. We are going to present them one by one.

We separate these operations in two main sets: productions and regulations.

Productions can take two different forms, depending on whether the reactants are consumed by the reactions or not:

- In Fig. 11.2, lactose and galactosidase produce glucose, and are consumed while doing so, which is thus noted (*galactosidase, lactose* ⇸ *glucose*).
- On the opposite, the expression of the lacZ gene to produce galactosidase (or of the lacl gene to produce the Lacl repressor protein) does not consume the gene, and we have thus (*lacZ* ⇸ *galactosidase*).

Generally speaking:

- If the reaction consumes completely the reactant(s) we write: $a_1, a_2, \ldots, a_n \to b$. Here the production of b completely consumes $a_1, \ldots, a_n$
- If the reactants are not completely consumed by the reaction, we write $a_1, a_2, \ldots, a_n \to b$. Here b is produced but $a_1, a_2, \ldots, a_n$ are still present after the production of b.

Regulations can also take two forms: every reaction can be either *inhibited* or *activated* by other proteins or conditions.

- In the example above, the production of galactosidase from the expression of the lacZ gene is activated by cAMP (we use *cAMP* ⇸ to express activation)
- At the same time the same production of galactosidase is blocked (or inhibited) by the Lacl repressor protein (noted *Repressor* ⊣).

Generally speaking:

---

[2]It is important here to notice that lactose can be either considered as a weak endogenous variable, or as an exogenous variable if we consider that the environment is always providing "enough" lactose. It is a simple example which shows that variables in a graph can be interpreted differently according to what is going to be observed.

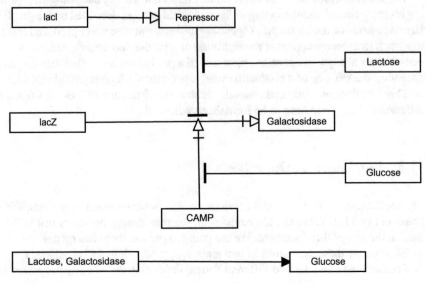

Fig. 11.2 Functional representation of the lac operon

- we write $a_1, a_2, ...a_n \rightarrow$ if the simultaneous presence of $a_1, a_2, ...a_n$ activates a production or another regulation.
- we write $a_1, a_2, ...a_n \dashv$ if the simultaneous presence of $a_1, a_2, ...a_n$ inhibits a production or another regulation.

On Fig. 11.3, we have a summary of basic inhibitions/activations on a reaction: the production of b from $a_1, ..., a_n$ is activated by the simultaneous presence of

Title: Activations and inhibitions

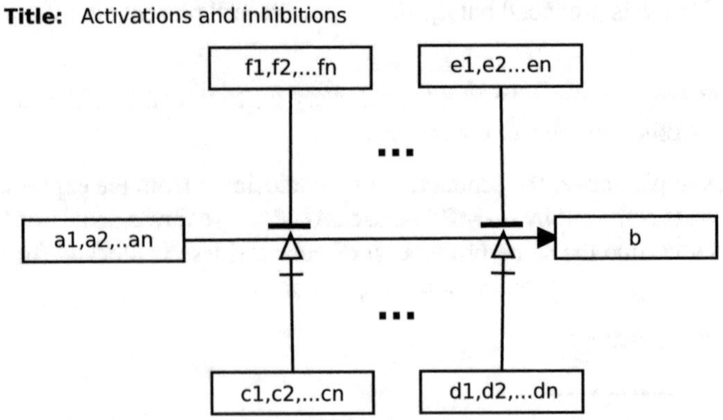

Fig. 11.3 Activations/Inhibitions

Title: Stacking regulations

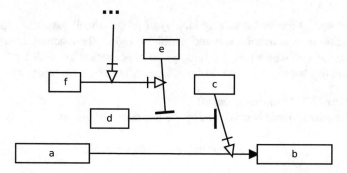

Fig. 11.4 Stacking

$c_1, \ldots, c_n$ **or** by the simultaneous presence of $d_1, \ldots, d_n$, and inhibited by the simultaneous presence of $e_1, \ldots, e_n$ **or** by the simultaneous presence of $f_1, \ldots, f_n$.

These regulations are often "stacked", on many levels (see Fig. 11.4). For example in Fig. 11.2, the inhibition by the Lacl repressor protein of the production of galactosidase can itself be inhibited by the presence of lactose, while the activation of the same production by cAMP is inhibited by the presence of glucose.

A final word of warning is necessary. Graphs pragmatically describe sequences of operations that biologists find important. They are only a model of some of the biological, molecular and chemical reactions that take place inside the cell; they can also be written in many different ways, depending on the functional block or operations that biologists want to describe, and some relationships are sometimes simply left out because they are considered not important for the function which is described in a particular graph.

11.4   Molecular Interaction Logic

In this section we extend a previous approach to logical modelling of graphs made in terms of first-order logic with equality [8–10]. Our approach, Molecular Interaction Logic (MIL), is based on modal temporal logic, which will help us later to define connections with other logical approaches to temporal reasoning as well as studying graphs behaviour in the context of a modal approach. We start this section by introducing the concepts of *pathway context* and *pathway formula*. The former corresponds to the formalization of *regulation* while the latter is the formal representation of the *production rules*, both concepts were presented in Sect. 11.3.

Definition 11.1 (*Pathway context*) Given a set of entities, a *pathway context* is formed by expressions defined by the following grammar:

$$\alpha ::= \langle \{\alpha_1, \cdots, \alpha_n\}P \twoheadrightarrow, \{\alpha_{n+1}, \cdots, \alpha_{n+m}\}Q \dashv \rangle$$

where P and Q are sets (possibly empty) of propositional variables representing the conditions of activation ($\twoheadrightarrow$) and inhibition ($\dashv$) of the reaction. Every context can be associated with a (possibly empty) set of activation (α_i, with $1 \le i \le n$) and inhibition (α_j, with $n < j \le m$) contexts. One, or both sets can be empty.   □

Definition 11.2 (*Pathway formula*)

A *Pathway formula* is generated by the following grammar:

$$F ::= [\alpha]\left(P^\wedge \multimap q\right) \mid F \wedge F$$

where α represents a context, $\multimap \in \{\twoheadrightarrow, \dashv\}$, $P^\wedge$ stands for a conjunction of all atoms in the set P and q corresponds to a propositional variable.   □

11.4.1 MIL Semantics

Before introducing the semantics we need to give a formal definition of the *activation* and *inhibition* expressions, since both concepts play an important role in the definition of the semantics.

Definition 11.3 (*Activation and inhibition expressions*)

Given a context of the form

$$\alpha = \langle \{\alpha_1, \cdots, \alpha_n\}P \twoheadrightarrow, \{\beta_{n+1}, \cdots, \beta_{n+m}\}Q \dashv \rangle,$$

we define the corresponding expressions $\mathcal{A}(\alpha)$ and $\mathcal{I}(\alpha)$ recursively as follows:

$$\mathcal{A}(\alpha) = \bigwedge_{p \in P} p \wedge \bigwedge_{i=1}^{n} \mathcal{A}(\alpha_i) \wedge \left(\bigvee_{q \in Q} \neg q \vee \bigwedge_{j=n+1}^{m} \mathcal{I}(\beta_j)\right)$$

$$\mathcal{I}(\alpha) = \bigvee_{p \in P} \neg p \vee \bigvee_{i=1}^{n} \mathcal{I}(\alpha_i) \vee \left(\bigwedge_{q \in Q} q \wedge \bigwedge_{j=n+1}^{m} \mathcal{A}(\beta_j)\right).$$

□

Informally speaking, $\mathcal{A}(\alpha)$ characterizes when the context α is active while $\mathcal{I}(\alpha)$ defines when it is inhibited. If one part of the context α is empty, then the corresponding part is of course absent in $\mathcal{A}(\alpha)$ and $\mathcal{I}(\alpha)$.

Definition 11.4 (*Extended signature*) Given a set of atoms Σ, its corresponding extended signature, $\widehat{\Sigma}$, is defined by the following expression:

$$\widehat{\Sigma} = \Sigma \cup \{\mathbf{Pr}\,(p) \mid p \in \Sigma\} \cup \{\mathbf{Cn}\,(p) \mid p \in \Sigma\},$$

where p is an endogenous variable.   □

Informally speaking, every atom of the form $\mathbf{Pr}\,(p)$ means that p is produced as a result of a chemical reaction. On the other hand, $\mathbf{Cn}\,(p)$ means that the reactive p has been consumed in a reaction. From now on, we will use the symbols Σ and $\widehat{\Sigma}$ referring to, respectively, the signature and its corresponding extension.

Definition 11.5 (*MIL interpretation*)
Let Σ be a set of propositional variables and $\widehat{\Sigma}$ its corresponding extended signature. We define a *MIL interpretation*, $\mathbf{V} = V_0, V_1, \ldots$, as an infinite sequence of sets of atoms on $\widehat{\Sigma}$ such that every endogenous variable $p \in \Sigma$ satisfies the following constraint:

$$\forall i \geq 0 \text{ if } \quad \mathbf{Pr}\,(p) \in V_i \text{ or } (\mathbf{Cn}\,(p) \notin V_i \text{ and } p \in V_i)$$
$$\text{then } \quad p \in V_{i+1}. \tag{11.1}$$

□

Definition 11.6 (*Satisfaction relation*) Given a MIL interpretation $\mathbf{V} = V_0, V_1, \ldots,$ $i \geq 0$ and a pathway formula F on Σ, we will define recursively the satisfaction relation $(\mathbf{V}, i \models F)$ as follows:

- $\mathbf{V}, i \models p$ iff $p \in V_i$, for any $p \in \Sigma$
- negation, disjunction and conjunction are satisfied as usual
- $\mathbf{V}, i \models [\alpha]\,(P^{\wedge} \rightarrow q)$ iff for all $j \geq i$, if $V, j \models \mathcal{A}(\alpha)$ and $P \subseteq V_j$, then $\mathbf{Pr}\,(q) \in V_j$ and for all $p \in P$, $\mathbf{Cn}\,(p) \in V_j$
- $\mathbf{V}, i \models [\alpha]\,(P^{\wedge} \rightarrow q)$ iff for all $j \geq i$ if $V, j \models \mathcal{A}(\alpha)$ and $P \subseteq V_j$ then $\mathbf{Pr}\,(q) \in V_j$. □

11.5 Translating Molecular Interaction Logic into Linear Time Temporal Logic

In this section, we consider the connection between Molecular Interaction Logic and Linear Time Temporal Logic (LTL) [19] by showing a translation from our formalism into a restricted subset of LTL in which only operators $\bigcirc$ and $\square$ are used. We start this section by providing some background on LTL.

Definition 11.7 (*Temporal language*) *Temporal formulas* are generated by the following grammar:

$$\varphi ::= \bot \mid p \mid \neg\varphi_1 \mid \varphi_1 \wedge \varphi_2 \mid \varphi_1 \vee \varphi_2 \mid \varphi_1 \rightarrow \varphi_2 \mid \bigcirc\varphi_1 \mid \square\varphi_1 \mid$$
$$\Diamond\varphi_1 \mid \varphi_1 \mathcal{U}\varphi_2 \tag{11.2}$$

where φ_1 and φ_2 are temporal formulas in their turn and p is any atom. Modal operators $\bigcirc$, $\square$, $\lozenge$ and $\mathcal{U}$ are respectively read as "next", "forever", "possible" and "until". $\square$

Definition 11.8 (*LTL semantics*) Let $\widehat{\Sigma}$ be a set of propositional variables. An LTL model is an infinite sequence, $\mathbf{V} = V_1, V_2, \ldots$, of sets of atoms on $\widehat{\Sigma}$. Given an LTL interpretation and $i \geq 0$, the LTL satisfaction relation is defined as follows:

1. $\mathbf{V}, i \models p$ iff $p \in V_i$, for $p \in \Sigma$.
2. Negation, conjunction and disjunction are satisfied in the usual way.
3. $\mathbf{V}, i \models \varphi \rightarrow \psi$ iff $\mathbf{V}, i \not\models \varphi$ or $\mathbf{V}, i \models \psi$.
4. $\mathbf{V}, i \models \bigcirc\varphi$ iff $\mathbf{V}, i + 1 \models \varphi$.
5. $\mathbf{V}, i \models \square\varphi$ iff for all $j \geq i$, $\mathbf{V}, j \models \varphi$.
6. $\mathbf{V}, i \models \lozenge\varphi$ iff there exists $j \geq i$, $\mathbf{V}, j \models \varphi$.
7. $\mathbf{V}, i \models \varphi\mathcal{U}\psi$ iff $\exists j \geq i$, $\mathbf{V}, j \models \psi$ and $\forall k$ s.t. $i \leq k < j$, $\mathbf{M}, k \models \varphi$.

$\square$

11.5.1 From MIL to LTL

Definition 11.9 (*Inertia rule*) Let p be an endogenous variable in a signature Σ. We define *inertia*(p) as the following formula built on $\widehat{\Sigma}$:

$$inertia(p) \stackrel{\text{def}}{=} \square((\mathbf{Pr}\,(p) \vee (p \wedge \neg\mathbf{Cn}\,(p))) \rightarrow \bigcirc p) \qquad (11.3)$$

$\square$

Thanks to these rules, we can specify how the truth values of biological substances evolve along time. More specifically, this rule means that a protein p might become true if it is the result of a production rule (concept represented by $\mathbf{Pr}\,(p)$) or if it is already present and it has not been used to produce other proteins (concept represented by $\mathbf{Cn}\,(p)$). By using structural induction we can prove the following proposition:

Proposition 11.1 *Let Σ be a finite signature. Given a LTL interpretation, $\mathbf{V}$, on $\widehat{\Sigma}$. $\mathbf{V}, 0 \models \bigwedge_p inertia(p)$, with p and endogenous variable in Σ, iff $\mathbf{V}$ satisfies condition (11.1) of Definition 11.5.* $\square$

Definition 11.10 (*Translation from MIL into LTL*) Let F be a pathway formula built on a signature Σ. We define the formula $tr\,(F)$, built on the signature $\widehat{\Sigma}$, as follows:

$$tr\left([\alpha]\left(P^{\wedge} \to q\right)\right) = \Box\left(\mathcal{A}(\alpha) \wedge P^{\wedge} \to \left(\mathbf{Pr}\,(q) \wedge \bigwedge_{p \in P} \mathbf{Cn}\,(p)\right)\right);$$

$$tr\left([\alpha]\left(P^{\wedge} \dashrightarrow q\right)\right) = \Box\left(\mathcal{A}(\alpha) \wedge P^{\wedge} \to \mathbf{Pr}\,(q)\right);$$

$$tr\left(F_1 \wedge F_2\right) = tr\left(F_1\right) \wedge tr\left(F_1\right),$$

where both F_1 and F_2 stand for two arbitrary pathway formulas. $\Box$

In order to guarantee that our translation is correct with respect to the MIL semantics presented in Sect. 11.4, we establish the following correspondence between both formalisms:

Lemma 11.1 (Correspondence) *Let F be a pathway formula built on a signature Σ, and **V** a LTL interpretation on the extended signature $\hat{\Sigma}$. Then we have the following equivalence:*

$$\mathbf{V}, 0 \models F \text{ iff } \mathbf{V}, 0 \models tr\,(F) \wedge \bigwedge_p inertia(p),$$

where p is an endogenous variable in F. $\Box$

11.6   Temporal Reasoning

As shown in the previous section, we can establish a correspondence between a graph and a temporal formula which describes its behaviour. However, in order to perform temporal reasoning we need to add the supplementary hypothesis of *closed world assumption*. This concept corresponds to the presumption that a statement that is true is also known to be true. Conversely, what is not currently known to be true, is false. This hypothesis fits perfectly in the biological process, as endogenous proteins appear if and only if a production rule is triggered (except at time 0 for weak endogenous variables) and, moreover, they are consumed only if they are used in a reaction.

11.6.1   Completion Axioms

In order to incorporate this hypothesis we define the *Completion axioms* [6] for our temporal theories as follows:

Definition 11.11 (*Completion axioms*) Let Σ be a finite signature and let F be a pathway formula built on Σ. For any pure endogenous propositional variable p in Σ, $COMP(F, p)$ corresponds to as the following formula built on $\hat{\Sigma}$:

$$COMP(F, p) = \neg p \wedge \Box(\bigcirc p \rightarrow (\mathbf{Pr}\,(p) \vee (p \wedge \neg \mathbf{Cn}\,(p))))$$

$$\wedge \Box \left(\mathbf{Pr}\,(p) \rightarrow \bigvee_{[\alpha](P^\wedge -\circ p)\in F} P^\wedge \wedge \mathcal{A}(\alpha) \right)$$

$$\wedge \Box \left(\mathbf{Cn}\,(p) \rightarrow \bigvee_{[\alpha](P^\wedge \rightarrow p)\in F} P^\wedge \wedge \mathcal{A}(\alpha) \right).$$

If p is a weak endogenous variable, $COMP(F, p)$ has the same form as above but omitting the conjunct $\neg p$. $\Box$

Broadly speaking, the meaning of the different components of $COMP(F, p)$ can be explained as follows:

- $\neg p$: this is a consequence of the type of the substance. If p is pure endogenous (it must be produced before existing), p must not be present at the initial state to that $\neg p$ must be part of $COMP(F, p)$. For the case of weak endogenous variables, whose truth value at the initial state cannot be deduced, requires that the conjunct $\neg p$ be omitted from corresponding completion formula.

- $\Box \left(\mathbf{Pr}\,(p) \rightarrow \bigvee_{[\alpha](P^\wedge -\circ p)\in F} P^\wedge \wedge \mathcal{A}(\alpha) \right)$: in any state, the production of a protein p is due to the satisfaction of, at least, one pathway formula.

- $\Box \left(\mathbf{Cn}\,(p) \rightarrow \bigvee_{[\alpha](P^\wedge \rightarrow p)\in F} P^\wedge \wedge \mathcal{A}(\alpha) \right)$: in any state, if p is consumed then it must be used in a reaction represented by a pathway formula.

- $\Box(\bigcirc p \rightarrow (\mathbf{Pr}\,(p) \vee (p \wedge \neg \mathbf{Cn}\,(p))))$: if a substance p is present then it has been produced in the previous state or it was already present and it was not consumed in a reaction.

If we consider now the whole set of propositional variables occurring in F, the resulting completion axioms correspond to the following conjunction

$$\bigwedge_p COMP(F, p), \text{ with } p \text{ being an endogenous variable.}$$

11.6.2 Graphs as Splittable Temporal Logic Programs

Completion axioms are used when we want to translate a non-monotonic theory into classical logic. To give an example, in the Answer Set Programming [4] paradigm, the answer sets of a propositional theory can be captured by a classical propositional expression by adding the so called *Loop formulas* [12, 18] (in the same spirit as Clark's completion). This result was extended to the case of non-monotonic temporal theories[3] in [2] in which it is shown that, regarding a syntactical restricted class of

---

[3]For a more detailed survey of temporal extension of Answer Set Programming see [1].

programs, called *splittable*, loop formulas can be effectively computed. We define such class of programs below:

Definition 11.12 (*Splittable temporal logic program*) A *splittable temporal logic program* Π for signature $\widehat{\Sigma}$ is said to be *splittable* if Π consists of rules of the form:

$$B^{\wedge} \wedge N^{\wedge} \to H \tag{11.4}$$

$$B^{\wedge} \wedge \bigcirc B'^{\wedge} \wedge N^{\wedge} \wedge \bigcirc N'^{\wedge} \to \bigcirc H' \tag{11.5}$$

$$\Box(B^{\wedge} \wedge \bigcirc B'^{\wedge} \wedge N^{\wedge} \wedge \bigcirc N'^{\wedge} \to \bigcirc H') \tag{11.6}$$

where B and B' are conjunctions of atoms, N and N' are conjunctions of negative literals like $\neg p$ with $p \in \widehat{\Sigma}$, and H and H' are disjunctions of atoms. $\qquad\square$

Roughly speaking, the idea behind a splittable program is that no past reference depends on the future.

Since the formalism presented in [2], called *Temporal Equilibrium Logic* (TEL) [1, 5], shares the syntax with LTL we can study our theories under such framework. As a result, we can translate, by using several temporal equivalences, our theories into splittable temporal logic programs, as stated in the following proposition:

Proposition 11.2 *Given a conjunction of pathway formulas* $F = F_1 \wedge \cdots \wedge F_n$, *it can be proved that*

$$tr(F) \wedge \bigwedge_{p} inertia(p),$$

where p corresponds to an endogenous variable in F, is equivalent to a splittable temporal logic program. $\qquad\square$

This equivalence allows us to study the relation between our completion axioms and loop formulas, which is considered next.

11.6.2.1 Relation with Loop Formulas

We have already shown that our temporal theories can be translated into splittable temporal logic programs under temporal equilibrium logic semantics. We now show how our completion axioms can be seen as a special case of loop formulas. Before presenting the result, we summarize how loop formulas are computed in [2]. Given a splittable program, Π, loop formulas are generated from the corresponding *(positive) dependency graph* of a temporal logic program Π, denoted by $G(\Pi)$. Nodes of $G(\Pi)$ correspond to the propositional variables in Π while edges are defined by the following expression:

$$E = \{(p, p) \mid p \in \Pi\} \cup \{(p, q) \mid \exists (B^{\wedge} \wedge N^{\wedge} \to p) \in \Pi \text{ s.t. } q \in B\}. \tag{11.7}$$

for any propositional variable p.

Definition 11.13 (*Loop from* [12]) A set of atoms L is called a loop of a logic program Π iff the sub-graph of $G(\Pi)$ induced by L is strongly connected. Notice that reflexivity of $G(\Pi)$ implies that for any atom p, the singleton $\{p\}$ is also a loop   □

When applying this technique to our translation (considering TEL semantics) we must consider the following points:

1. Given a conjunction of pathway formulas F, $tr(F)$ has no positive cycles in the sense of [2]. This means that only unitary cycles must be considered.
2. The hypothesis of closed world assumption should not be applied to the exogenous variables, whose presence cannot be justified and whose absence cannot be determined "by default". They must remain *free*, specially when querying our representation.

Item 1 means that the computation of the loop formulas, as presented in [2], is equivalent to our completions axioms (see Definition 11.11), while 2 means that completion rules should not be computed in the case of exogenous variables. This result is stated in the following proposition:

Proposition 11.3 *Completion axioms of Definition 11.11 are equivalent to loop formulas (under TEL semantics) when they are restricted to endogenous variables (a concept explained in Sect. 11.2).*   □

This result justifies that our approach can be also considered as non-monotonic temporal logic programs.

11.6.3 Grounding Splittable Temporal Logic Programs

The use of an LTL formalization allows us to consider solutions with infinite length when performing reasoning tasks such as abduction or satisfiability. However, regarding complexity results, it is worth to mention that LTL satisfiability is, in the general case PSPACE-complete while, regarding the propositional case, it is *NP*-complete. In an attempt to reduce the complexity of the problem as well as taking advantage of the tools available for reasoning on propositional logic such as SAT-solvers, abduction algorithms, etc., we consider *bounded time*, that is, we fix the positive constant *max* as the maximum time length. This assumption allows us to translate the temporal formulas into a propositional theory, as explained below.

Definition 11.14 Let φ a temporal formula built on the language presented in (11.2), $max \geq 0$ and $0 \leq i < max$. We define translation of φ, at instant i, into propositional logic, denoted by $\langle \varphi \rangle_i$, as follows:

- $\langle p \rangle_i \overset{\text{def}}{=} p_i$, with p an atom and p_i a new propositional variable;
- $\langle \neg \varphi \rangle_i \overset{\text{def}}{=} \neg \langle \varphi \rangle_i$;

- $\langle \varphi \odot \psi \rangle_i \overset{\text{def}}{=} \langle \varphi \rangle_i \odot \langle \psi \rangle_i$, with $\odot \in \{\wedge, \vee, \rightarrow\}$;
- $\langle \bigcirc \varphi \rangle_i \overset{\text{def}}{=} \langle \varphi \rangle_{i+1}$;
- $\langle \Box \varphi \rangle_i \overset{\text{def}}{=} \bigwedge_{i \leq j < max} \langle \varphi \rangle_j$;
- $\langle \Diamond \varphi \rangle_i \overset{\text{def}}{=} \bigvee_{i \leq j < max} \langle \varphi \rangle_j$;
- $\langle \varphi \mathcal{U} \psi \rangle_i \overset{\text{def}}{=} \bigvee_{i \leq j < max} \left(\langle \psi \rangle_j \wedge \bigwedge_{i \leq k < j} \langle \varphi \rangle_k \right).$

$\Box$

Broadly speaking, this translation simulates the truth value of an LTL propositional variable p along time by a set of n fresh atoms in classical logic, one per time instant. Moreover, the behaviour of modal operators are simulated by (finite) conjunctions and disjunctions, since we are considering bounded time. The following observation shows that, under the assumption of bounded time, we can establish a one-to-one correspondence between temporal and grounded theories.

Observation 11.1 (Model correspondence) *Let* $\mathbf{V} = V_0, V_1, \ldots$ *be an LTL interpretation. Given* $max \geq 0$, *we define the classical interpretation* I_{max} *as:*

$$I_{max} = \{p_i | p \in V_i\}.$$

It can be proved that $\mathbf{V}$ *and* I_{max} *satisfy the following property:*

$$\forall \varphi, \ \mathbf{V}, i \models \varphi \ \text{iff} \ I_{max} \models \langle \varphi \rangle_i.$$

$\Box$

11.7 Reasoning and Solving

In the previous section we described the theoretical aspects of the representation of graphs and of the logic used for reasoning on them.

In this section, we are going to show that, after translation, any question can be expressed in classical propositional logic and solved using classical propositional logic tools and that even complex questions, such as the search for a stable state, can be solved by our system.

11.7.1 A Simple Example

We are first going to explain on a simple example how the transformation of a graph into a set of CNF formulas is performed. We are going to work on the graph describing

Title: lac operon regulation simplified

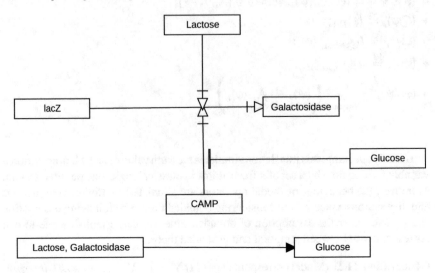

Fig. 11.5 Simplified functional representation of the lac operon

the behaviour of the Lac operon represented on Fig. 11.2 in Sect. 11.2. However we simplify a little this graph into the one represented in Fig. 11.5.

In this example, *lacZ* and *cAMP* are *exogenous* variables. As their value is set once and for all, they don't have to be grounded in the translation. On the opposite, *Galactosidase* is a *pure endogenous* variable, and thus will be grounded. *Glucose* is a *weak endogenous* entity: it has to be grounded as its value can change through time, however no completion formula will be computed for the variable describing them at time 0, as they can be present at the start of the process as an initial condition. Here, we will consider *Lactose* as an *exogenous* variable (see footnote 2 in Sect. 11.2).

This graph is interesting, because it has a temporal dynamic. For example, if the initial conditions are that *lacZ*, *cAMP* and *Lactose* are present and *Glucose* is absent then we can simulate informally the evolution of the proteins/conditions as:

Time 0:   *lacZ, cAMP, Lactose*
Time 1:   *lacZ, cAMP, Lactose, Galactosidase*
Time 2:   *lacZ, cAMP, Lactose, Galactosidase, Glucose*
Time 3:   *lacZ, cAMP, Lactose, Glucose*

Time 3 is a stable state.

Now we are going to see how this informal process can be formalized, and how logical tools can be used to reason about this graph; in a first step, this graph can be represented by:

$$Lactose \wedge Galac_t \rightarrow pr(Glucose)_t \qquad (11.8)$$

$$lacZ \wedge Lactose \wedge cAMP \wedge \neg Glucose_t \rightarrow pr(Galac)_t \qquad (11.9)$$

$$Lactose \wedge Galac_t \rightarrow cn(Galac)_t \qquad (11.10)$$

$$pr(Glucose)_t \rightarrow Glucose_{t+1} \qquad (11.11)$$

$$pr(Galac)_t \rightarrow Galac_{t+1} \qquad (11.12)$$

$$Glucose_t \wedge \neg cn(Glucose)_t \rightarrow Glucose_{t+1} \qquad (11.13)$$

$$Galac_t \wedge \neg cn(Galac)_t \rightarrow Galac_{t+1} \qquad (11.14)$$

Equations 11.8 and 11.9 describe how proteins can be produced: *Glucose* is produced (at time t) when we have *Lactose* and *Galac* at time t (lactose is always present as we suppose that there is always enough lactose in our environment, while galactosidase evolve in time), and *Galac* is produced at t when we have *Glucose* at t along with *cAMP*, *Lactose* and the *lacZ* gene.

Equation 11.10 expresses that when we have *Lactose* and *Galac* at t then *Galac* is consumed at t as it is used to produce *Glucose* according to Eq. 11.8. We have no similar equation for *Lactose*, as *Lactose* is exogenous and there will always remain "enough" lactose.

Equations 11.11 and 11.12 express that, when a molecule is produced at time t, then it is present at time $t + 1$. This applies here to *Glucose* and *Galac*.

Equations 11.13 and 11.14 are inertia rules. If a protein is present at time t and is not consumed at t then it will be present at time $t + 1$.

After completing this first representation, the system has to be grounded by time. The number of time steps is chosen by the user. Grounding is trivial and, for one time step, the above set of rules just becomes:

$$Lactose \wedge Galac_0 \rightarrow pr(Glucose)_0$$

$$lacZ \wedge Lactose \wedge cAMP \wedge \neg Glucose_0 \rightarrow pr(Galac)_0$$

$$Lactose \wedge Galac_0 \rightarrow cn(Galac)_0$$

$$pr(Glucose)_0 \rightarrow Glucose_1$$

$$pr(Galac)_0 \rightarrow Galac_1$$

$$Glucose_0 \wedge \neg cn(Glucose)_0 \rightarrow Glucose_1$$

$$Galac_0 \wedge \neg cn(Galac)_0 \rightarrow Galac_1$$

Then, we build completion formulas. It is important to notice that completion formulas are *always* built for *pure endogenous* variables at all time steps, are *never* built for *exogenous* variables, and are built for *weak endogenous* variables at all time steps *except* at time 0. This is a consequence of the "closed world" assumption: pure endogenous entities can only be created, produced or consumed internally, and are never present at the start of the process. So, to take a simple example, each time we have multiple paths to produce a pure endogenous variable p, such as $C_1 \rightarrow Pr(p)$ up to $C_n \rightarrow Pr(p)$, then we must add a "completion" formula $Pr(p) \rightarrow C_1 \vee \cdots \vee C_n$.

The following completion formulas are respectively associated with the variables:
(1) $Galac_0$, (2) $Galac_1$, (3) $Glucose_1$, (4) $cn(Galac)_0$, (5) $pr(Galac)_0$,
(6) $pr(Glucose)_0$.

$$\neg Galac_0 \tag{11.15}$$
$$Galac_1 \rightarrow (Galac_0 \wedge \neg cn(Galac)_0) \vee pr(Galac)_0 \tag{11.16}$$
$$Glucose_1 \rightarrow (Glucose_0 \wedge \neg cn(Glucose)_0) \vee pr(Glucose)_0 \tag{11.17}$$
$$cn(Galac)_0 \rightarrow (Galac_0 \wedge Lactose) \tag{11.18}$$
$$pr(Galac)_0 \rightarrow (lacZ \wedge Lactose \wedge cAMP \wedge \neg Glucose_0) \tag{11.19}$$
$$pr(Glucose)_0 \rightarrow (Galac_0 \wedge Lactose) \tag{11.20}$$

Then, all formulas are automatically translated into CNF. With one temporal grounding step, the simple graph considered here is represented by a database of 21 CNF formulas.

11.7.2 From Temporal Reasoning to Classical Propositional Tools

As a graph is now transformed into a database D of propositional CNF formulas, any propositional tool can be used to solve queries.

Some questions Q such as *"is molecule p present at time 3"* can be expressed by the logical temporal formula $\bigcirc \bigcirc \bigcirc p$, and then easily translated after grounding into the classical p_3. Then it can be solved with a SAT solver: $\neg Q = \neg p_3$ is added to D and the satisfiability of $D \cup \neg Q$ is checked. If it is not satisfiable, then Q is of course true.

However, most often, the main problem for biologists is to find the set(s) of conditions/preconditions that will lead to the creation of a protein, or the triggering of a specific condition. For example, many graphs describe how some cellular paths lead to cell death (apoptosis). Then the question is usually "what are the set(s) of condition(s) that lead to cell apoptosis after some time". Here depending on the complexity of the problem, different tools can be used:

- Abduction is of course the more natural and elegant solution. If we call D the set of CNF formulas after grounding into propositional logic, and Q the question, then we first use a SAT solver to check that $D \cup \{\neg Q\}$ is consistent (if it is not consistent then Q is already an implicate of D). Now we search for the minimal set H such as Q is an implicate of $T \cup H$. The classical algorithm consists in computing the set $\mathcal{P}(D \cup \{\neg Q\})$, which is the set of the prime implicates (the strongest clausal consequences) of $D \cup \{\neg Q\}$, and then checking for each $x \in \mathcal{P}(D \cup \{\neg Q\})$ that $D \cup \{\neg x\}$ is consistent. Then each such x is a solution.

While solutions sets containing only exogenous variables and weak endogenous variables (at step 0) describe the initial conditions leading to Q, abduction is able to find *all* sets answering a given question, even sets containing pure endogenous variables. This can give valuable information regarding the cell internal dynamic.

• While abduction is the more elegant way to find the set of preconditions answering a given question, the number of prime implicates of a theory can be exponential in the size of the theory and finding only one implicate is an NP-hard problem [13]. Thus the underlying complexity when using large graphs may turn abduction into an impracticable method and an alternative approach has to be used.

Biologists are mainly interested in solutions that contain only *exogenous* variables and also initial conditions which are the values of *weak endogenous* variables at time step 0. As explained before, *exogenous* proteins are interesting candidates as they usually describe the external conditions that can be set to activate some specific paths inside the graph, and *weak endogenous* variables at step 0 describe initial conditions. *Pure endogenous* proteins only depend on the internal dynamic of the cell.

If we call $Ex0$ the set of exogenous variables and weak endogenous variables at step 0, an extensive search can be performed with a Sat Solver to check if there is a valuation satisfying $D \cup \{\neg Q\}$ for a fixed boolean affectation of the set $Ex0$. If so, then each such affectation is a solution.

This method is faster than abduction as long as the set of exogenous variables/weak endogenous variables at step 0 remains small. The complexity however grows exponentially with the number of variables in the set, and it can't provide the solution sets containing *pure endogenous* variables, nor weak endogenous variables at a time greater than 0.

11.7.3 Expressing Complex Queries

In the previous section we presented the reasoning tools that can be used to answer simple questions. In this section we show how much more complex questions can be expressed and solved.

Any question that can be expressed using the temporal logic described in Sect. 11.5 can be solved using classical propositional reasoning tools, as it can be translated into propositional logic (considering of course bounded time). For example, if we want to know if the introduction of protein p will produce protein q at time 3, we just have to solve the question $Q = p \rightarrow q_3$, i.e. check if $D \cup \{p\} \cup \{\neg q_3\}$ is inconsistent.

More complicated, questions can be asked. For example, if we want to know if a stable state exists, we just write:

$$Q = \Diamond\Box \bigwedge_{p \in En} (p \leftrightarrow \bigcirc p)$$

where *En* is the set of endogenous variables (pure and weak). The value of exogenous variables never change, so they are always "stable". This is grounded and translated into propositional logic as

$$Q = \bigvee_{0 \le i \le n} \bigwedge_{i \le j \le n} \bigwedge_{p \in En} (p_j \leftrightarrow p_{j+1})$$

where n is the last grounding step. We then add $\neg Q$ to D and check if $D \cup \{\neg Q\}$ is inconsistent.

Having the full expressivity of temporal logic to write queries is an important feature of our system. Users are able to build complex queries, and they can be automatically translated and solved by the system. Solving the question can be a "yes/no" answer using a simple consistency check, or a more elaborate answer which will provide the set(s) of conditions which lead to a "yes" answer, using abduction or using the exhaustive search method described in the previous section.

11.8 Implementation

We have already implemented most of the tools necessary for using the system:

- Graphs are built using *Pathvisio* [20], a public-domain editing software and a well known tool in the biologists community.
- We have developed a parser/translator which reads the XML files generated by Pathvisio, takes as an argument the number of grounding steps, and translates the graphs into a set of classical grounded propositional CNF formulas.
- For consistency check and exhaustive search, we use the Glucose SAT solver [3] which is based on Minisat [11]. To compute prime implicates, we implemented our own version of the Tsiknis, Dean and Johnson algorithm [14, 15, 17]. While our implementation, which is based on machine language operations, seems to be extremely fast, a more exhaustive comparison with other approaches for computing prime implicants,[4] such as the ones advocated in [7, 13], should be tested.

11.9 Conclusion

We have presented in this paper a method to translate graphs representing biological systems into temporal logic formulas and to solve complex temporal queries regarding these graphs. This method has been almost fully implemented, and the associated tool has now reached a state where it can be tested on large, realistic graphs.

---

[4]The dual problem, which could be easily adapted to suit our needs.

There remains however different points to address:

- Currently the graphs we are using are limited: they can only use elementary relations; while all existing relations can be expressed with this elementary subset, it would be easier (and would keep graphs smaller) if our graph editor and out parser could deal with a larger subset of f these relations.
- graphs are built by hand by biologists, and they very often rely on "common knowledge" among them, so they sometimes "forget" to write some relations or sometimes express some relations between proteins in a non "standard" way. This tool will detect such missing knowledge and will thus help in writing more complete and consistent graphs, but correcting these problems is a mandatory step.
- While translating temporal logic queries into grounded propositional CNF is a technicality, building and understanding the exact meaning of a temporal query is complicated for people who don't have a training in logic. The goal of our users is to solve problems related to these graphs and we have to be able to describe the reasoning tasks available in a simple way, and give simple tools to write queries that can be solved by our system. A possible solution would be to provide a graphical interface that would help building queries by assembling intuitively variables and connectors as an intermediate between logic and natural language.
- Our system relies on a strong assumption: proteins can either be present or absent, but we are not able to consider partial concentrations. This decision was discussed with the biologists, and they supported it for a simple reason: currently, they are most of the time, if not all of the time, enable to determine the concentration of proteins in a cell. Their understanding of the cell chemical reactions is not precise enough, and they really consider the concepts of "absence", "presence", "Production" or "Consumption" when building graphs. However, this does not mean that we will not have to deal with this problem in the future.

The fact that there is now a demand from the biologists we are working with to get the tool and use it by themselves seems to prove that it has reached a certain state of maturity and stability, even if there probably remains work to do before turning it into a fully operational tool.

Acknowledgments This work is partially supported by ANR-11-LABX-0040-CIMI within the program ANR-11-IDEX-0002-02, by IREP Associated European Laboratory and by project CLE from Région Midi-Pyrénées.

References

1. Aguado, F., Cabalar, P., Diéguez, M., Pérez, G., Vidal, C.: Temporal equilibrium logic: a survey. J. Appl. Non-Class. Logics **23**(1–2), 2–24 (2013)
2. Aguado, F., Cabalar, P., Pérez, G., Vidal, C.: Loop formulas for splitable temporal logic programs. In: Proceedings of the 11th International Conference on Logic Programming and Non-monotonic Reasoning (LPNMR'11), pp. 80–92. Vancouver, Canada (2011)

3. Audemard, G., Simon, L.: Predicting learnt clauses quality in modern sat solver. In: Proceedings of the Twenty-First International Joint Conference on Artificial Intelligence (IJCAI'09), pp. 399–404 (2009)
4. Brewka, G., Eiter, T., Truszczyński, M.: Answer set programming at a glance. Commun. ACM **54**(12), 92–103 (2011)
5. Cabalar, P., Pérez, G.: Temporal equilibrium logic: a first approach. In: Proceedings of the 11th International Conference on Computer Aided Systems Theory (EUROCAST'07), pp. 241–248 (2007)
6. Clark, K.L.: Negation as failure. In: Logic and Databases, pp. 293–322. Plenum Press (1978)
7. Déharbe, D., Fontaine, P., LeBerre, D., Mazure, B.: Computing prime implicants. In: Formal Methods in Computer-Aided Design (FMCAD), pp. 46–52. Portland, USA (2013)
8. Demolombe, R., Fariñas del Cerro, L., Obeid, N.: Automated reasoning in metabolic networks with inhibition. In: 13th International Conference of the Italian Association for Artificial Intelligence, AI*IA'13, pp. 37–47. Turin, Italy (2013)
9. Demolombe, R., Fariñas del Cerro, L., Obeid, N.: Logical model for molecular interactions maps. In: Fariñas del Cerro, L., Inoue, K. (eds.) Logical Modeling of Biological Systems, pp. 93–123. Wiley (2014)
10. Demolombe, R., Fariñas del Cerro, L., Obeid, N.: Translation of first order formulas into ground formulas via a completion theory. J. Appl. Logic **15**, 130–149 (2016)
11. Een, N., Sorensson, N.: An extensible sat-solver. In: Proceedings of the 6th International Conference on Theory and Applications of Satisfiability Testing (SAT2003), pp. 502–518. Santa Margherita Ligure, Italy (2003)
12. Ferraris, P., Lee, J., Lifschitz, V.: A generalization of the lin-zhao theorem. Ann. Math. Artif. Intell. **47**(1–2), 79–101 (2006)
13. Jabbour, S., Marques-Silva, J., Sais, L., Salhi, Y.: Enumerating prime implicants of propositional formulae in conjunctive normal form. In: Proceedings of the 14th European Conference, JELIA 2014, pp. 152–165. Funchal, Madeira, Portugal (2014)
14. Jackson, P.: Computing prime implicates. In: Proceedings of the 20th ACM Conference on Annual Computer Science (CSC'92), pp. 65–72. Kansas City, USA (1992)
15. Jackson, P.: Computing prime implicates incrementally. In: Proceedings of the 11th International Conference on Automated Deduction (CADE'11), pp. 253–267. Saratoga Springs, NY, USA (1992)
16. Jacob, F., Monod, J.: Genetic regulatory mechanisms in the synthesis of proteins. J. Mol. Biol. **3**, 318–356 (1961)
17. Kean, A., Tsiknis, G.: An incremental method for generating prime implicants/implicates. J. Symbolic Comput. **9**, 185–206 (1990)
18. Lin, F., Zhao, Y.: ASSAT: computing answer sets of a logic program by sat solvers. In: Artificial Intelligence, pp. 112–117 (2002)
19. Pnueli, A.: The temporal logic of programs. In: Proceedings of the 18th Annual Symposium on Foundations of Computer Science, pp. 46–57. Providence, Rhode Island, USA (1977)
20. van Iersel, M.P., Kelder, T., Pico, A.R., Hanspers, K., Coort, S., Conklin, B.R., Evelo, C.: Presenting and exploring biological pathways with PathVisio. BMC Bioinform. **9**, 399 (2008)
21. Wikipedia: The lac operon. https://en.wikipedia.org/wiki/Lac_operon (2015)

Chapter 12
Jair Minoro Abe on Paraconsistent Engineering

Seiki Akama

Dedicated to Jair Minoro Abe for his 60th birthday

Abstract An overview of Professor Abe's scientific work is presented, emphasizing the main results obtained by him in his research activity. He has done a lot of work on paraconsistent logics and their applications. We survey his academic career and published works.

Keywords Jair Minoro Abe · Paraconsistent logics · Annotated logics

12.1 Introduction

Jair Minoro Abe has established itself as one of leading figure in the consolidation regarding to applications of paraconsistent systems, which had one of the introducers the renowned Prof. Newton C.A. da Costa. Abe has dedicated to an important class of paraconsistent logic, namely the paraconsistent annotated logics.

This paper intends to give a short view of activities of Abe. In section, we give his biographical information. In Sect. 12.3, we present a general description of his published works.

S. Akama (✉)
C-Republic, 1-20-1 Higashi-Yurigaoka, Asao-ku, Kawasaki 215-0012, Japan
e-mail: akama@jcom.home.ne.jp

© Springer International Publishing Switzerland 2016
S. Akama (ed.), *Towards Paraconsistent Engineering*, Intelligent Systems Reference Library 110, DOI 10.1007/978-3-319-40418-9_12

227

12.2 Biographical Information

Jair Minoro Abe was born on October 6, 1955 in São Paulo City, São Paulo, Brazil, as an eldest son of Tadashi Abe (1923–1984) and Kinuko Abe (1931-), typical Japanese immigrants from the 30s of last century, has two sisters, Marina and Nilza. Abe is married with Tiyo and has two daughters, Clarissa and Letićia.

By the age of four lived with uncles in Mogi das Cruzes city, nearby from São Paulo, where he studied Japanese language, not continuously until late 1962 due health conditions. He went back to São Paulo to start primary school (Grupo Escolar de Vila Ré) until mid 1965 when he moved to a newer building that replaced the old wooden shed and was named "Grupo Escolar Prof. Jose Bartocci". Abe studied there until 1966, the year that also attended "Externato Cristo Rei" to take the exam to enter to the "Colégio Estadual Prof. Gabriel Ortiz" (junior high school and high school) during the period 1967–1970 and 1971–1973.

Abe mentions that in primarily and high school times experienced one of the most amazing periods of his life either learning, but also meeting teachers and wonderful friends who greatly influenced the subsequent journey. Also in 1973 he attended the "CECA Vestibulares" by action of Prof. Julio Takara electing him as one of the best students, gracing him with a scholarship for preparatory course, which was of great importance for Abe at the time.

In 1974, he began the course of Bachelor of Mathematics at Institute of Mathematics and Statistics, University of São Paulo (USP). He had contact with valuable mathematicians such as M. Peixoto, E. Farah, N. da Costa, L. Berthet, J. Zimbarg, C. Hönig, O. Alas, A. Gillioli, among others. Shortly after completion, began the Graduate course in Pure Mathematics at same Institute having the supervision of Dr. Newton Costa and wrote his dissertation on foundations of ordered geometry [1].

After completing the master course, he attended a doctorate course at the Faculty of Philosophy, Letters and Human Sciences of University of São Paulo, where he obtained his Doctor degree in Philosophy under the supervision of Dr. Newton Costa [3]. Abe investigated the foundations of annotated logics, an issue that came to the hands of his mentor through an application in logic programming made by Blair and Subrahmanian [19, 32].

The appearance of a paraconsistent system in logic programming would provide the opening for applications, awaited the paraconsistent systems. Abe then had long conversations with Newton Costa on applications of paraconsistent systems and then designed a program to do it. At the time, invited to give a graduate course at Polytechnic School, University of São Paulo with Newton Costa, brought together some disciples to accomplish his project.

By 1996 Abe together with Prado and Avila implemented a logic programming language *Paralog* based on annotated logics [18, 22] independently of Subrahmanian and thus originated the first incursions in Artificial Intelligence (AI): an architecture based entirely on annotated logics and knowledge representation theory via the concept of frames [2].

A few years later, Da Silva Filho built electronic circuits accommodating inconsistencies. Among contributions in his doctoral thesis was included logic controller based on annotated logics called Para-analyzer. It was soon materialized into a logic controller called paracontrol. To accomplish its functionality it was built the first robot made entirely with hardware based on such logic: Emmy. There have been made many improvements.

Da Silva Filho noted that a convenient combination of algorithm Para-analyzer resulted in a 'network' which was named paraconsistent artificial neural network. Effectively it was verified that it had characteristics of an artificial neural network; furthermore such network present useful properties that differ much from existing ones.

Meanwhile, Abe felt the need to broaden the horizons of their research as well as for his collaborators: a feat that direction was the development of an entirely geared congress devoted to applications of logic to AI and technology. Thus the Congress of Logic Applied to technology (LAPTEC) was born. LAPTEC was welcomed with great enthusiasm and had some famous lecturers: P. Suppes, N. da Costa, E.G.K. Lopez-Escobar, M.C. Monard, N. Ebecken, K. Nakamatsu, S. Akama, T. Date, D. Dubois, E. Massad, M. Droste, and others.

At the 1st Congress of Paraconsistency held in Ghent, Belgium, 1997, Abe met K. Nakamatsu and S. Akama for the first time. They were studying annotated logics among their themes and expected to meet with Abe and was, indeed, an important milestone. They then proceeded to have a strong cooperation that last until nowadays.

Abe's carrier was made mainly hard work, but he thinks that only this it is not enough; it is necessary something more, that he call 'fortunate'. Abe also mention that he always had the support of his parents and after the death of his father, Abe continued to live whenever possible with his mother who gave him all the necessary support for the day-by-day of his career.

Abe's academic positions are as follows:

- Assistant Professor, Paulista State University, 1984–1995
- Coordinator of Logic and Science Theory, Institute For Advanced Studies, University of São Paulo, 1987–2016
- Research Associate, University of São Paulo, 1989–2016
- Full Professor, Paulista University, 1996-

All academic activities were done as Full Professor at Paulista University, which has received great deal to accomplish his investigations. Abe supervised many Ph.D. and M.Sc. students; see Ávila [17], Prado [31], Da Silva Filho [24]. Ph.D. students are as follows:
    Ph.D. students

- Bráulio Coelho Avila (Computational Intelligence)
- José Pacheco de Almeida Prado (Computational Intelligence)
- João Inácio da Silva Filho (Automation and Robotics)
- João Carlos Almeida Prado (Computational Intelligence)

- Mauricio Conceicão Mário (Computational Intelligence)
- Cláudio Rodrigo Torres (Automation and Robotics)
- Marcelo Nogueira (Computational Intelligence)
- Fábio Vieira do Amaral (Computational Intelligence)
- Nélio Fernando dos Reis (Decision-Making)
- Cristina Corrêa de Oliveira (Computational Intelligence)
- Avelino Palma Pimenta Jr (Computational Intelligence)

Abe's undergraduate courses were or are all introductory, having no prerequisites and presupposing no previous knowledge. In each course he takes care making the subject as attractive as possible with applications and/or possible applications.

Many of classes he used to teach playfully, explaining as detailed as possible. Stressing the priority of education, Abe strives to assist his students to think by themselves, that logic is a wonderful tool to do that, to gain independence of judgment. Besides their classes were really good: once da Costa wrote in one of his letters of recommendation that Abe was an excellent expositor. Also many of his students refer to him as master who transform difficult topics to understandable ones.

Among courses that Abe has lectured trough years are: differential calculus, basic algebra, linear algebra, numerical analysis, basic logic, basic non-classical logic, introduction to set theory, matrix theory, basic mathematics, basic statistics, vectors and geometry, computability theory, discrete mathematics, artificial intelligence, intelligent information systems, among others.

Among graduate courses, Abe has lectured: expert systems in production engineering, quantitative method in engineering, introduction to set theory, AI in bioinformatics, introduction to classical logic, introduction to non-classical logic, among others.

Abe organized (or co-organized) several international conferences including *Logic Applied to Technology* (LAPTEC) in 2000, 2001, 2002, 2003, 2005, 2007 (with J.I. da Silva Filho in 2005, K. Nakamatsu in 2007) and *Workshop Intelligent Computing Systems* (WICS) in 2013, 2014, 2015.

He also served as a reviewer for scientific journals including *Mathematica Japonica* (Editorial Board), *Scientiae Mathematicae Japonicae* (Editorial Board), *International Journal of Reasoning-based Intelligent Systems* (Advisory Editor), *Neurocomputing*, *Mathematical Reviews*.

12.3   General Description of Published Works

In this section, we briefly describe Abe's work. He has engaged in the following principal areas of research, namely:

1. Annotated logics
2. Artificial neural networks
3. Expert systems in decision-making

4. Automation and robotics
5. Curry algebras
6. Annotated systems and fuzzy set theory
7. Nelson logics
8. Annotated modal systems
9. Annotated logic programming
10. Logic and biology
11. Logic and psychoanalysis

As above, Abe studied many applications of paraconsistent logics to several areas. For a survey on applications to AI, see Abe [5].

Abe undoubtedly made important contributions to *annotated logics* [3, 21, 23], which belong to paraconsistent logics. He was to carry out a systematic study of such logics and was the first to write a dissertation on annotated logics. He established foundations for annotated logics like basic theory of models, including the Łoś theorem [3, 9, 10]. Algebraic versions were also investigated providing, in particular, completeness and decidability theorems.

For *artificial neural networks* [7] and their applications, he started with da Silva Filho, and students, applying in aiding of Alzheimer's disease diagnosis [26], the craniometric variables analysis [27], in speech disorder, typed characters recognition, and other issues [8].

For *expert systems in decision-making*, Abe devoted considerably in implementing the annotated evidential $E\tau$ paraconsistent logic in the matter of decision-making applied by innumerous MSc students and Ph.D. dissertations [20].

Abe also worked on *automation and robotics* based on paraconsistent logics. Namely, he developed with his students multiple robots resulted from the application of logical controller *Paracontrol*: Emmy, Sofya, Amanda, Hephaestus, all of them by using sensors of different types in order to and an electronic device for visual and/or hearing impaired who named Keller [25].

Abe has applied the concept of *Curry algebras* [6] in order to obtain algebraic versions not only for annotated logics, but also other class of paraconsistent and paracomplete systems. Abe has extended to first order monadic calculi of such systems via ideas of Halmos concerning monadic algebras [12].

There are many ways to obtain annotated set theories. One way to do is "inside" of some usual set theory (for instance, ZF-set theory) exactly as classical Fuzzy set theory as did by Zadeh. Abe has studied in this direction and one its versions (annotated set theory) encompasses Fuzzy Set theory [3, 21]. In collaboration to S. Akama, it was possible to adapt annotated axiomatic to obtain some axiomatizations of versions of fuzzy systems, showing the power of these systems [15].

Abe coauthored some papers with Akama in elucidating the operator "negation" in several non-classical logics like *Nelson logics* which have been developed as constructable systems by Nelson [30]. Akama, Abe and Nakamatsu proposed constructive discursive logics in [16].

Annotated modal systems can provide the basis for the paraconsistent, paracomplete and non-alethic reasoning, non-monotonic reasoning, defeasible reasoning, deontic reasoning, other doxastic logics, temporal logics, muti-modal logics, among others [4, 11, 14].

Nakamatsu and Abe have organized several invited sessions of several conferences. Abe has also participated actively with Nakamatsu in his research themes that lean on annotated logic programming, including defesiable deontic control systems [28, 29].

With the renowned entomologist N. Papavero, who became interested in the axiomatization of biology, Abe worked in collaboration on the theme in various aspects: firstly they considered Mereology as the basis of the issue, having as primitive concept "is part of" (e.g., 'the arm is part of the body'). Papavero and Abe also considered set theoretical predicates for axiomatization (in the sense of Suppes [33]) and they have succeed in Cladistics, in the conception of W. Hennig.

Abe has given assistance for the study of Lacan's proposal, aiding in the logical concepts used on his books of the Seminar.

For books, Abe, Akama and Nakamatsu published a book "Introduction to Annotated Logics" in 2015, which describes the theoretical basis of annotated logics [13]. In the same year, Abe also edited a book "Paraconsitent Intelligent Based-Systems" [8] entirely devoted to the applications of annotated systems. Abe, Akama and Nakamatsu plan to write more books on application of paraconsistent systems.

Outside the academic sphere, Abe want mention his taste for classical music and popular music. Like so many of his generation was influenced by various types of music of his time, but it highlights his taste for American singer Johnny Mathis, bossa nova rhythm, and pop music is 60s and oldies.

Abe also always liked pets (he'd had many of them through his life) that helped to distract in his spare time and also cultivated photographs, interest in past times events. Also he likes play regularly tennis and he regularly monitors major tournaments.

Acknowledgments I am grateful to Prof. Jair Minoro Abe for his valuable comments.

References

1. Abe, J.M.: Fundamentos da Geometria Ordenada (in Portuguese). MSc Thesis, University of São Paulo, São Paulo (1983)
2. Abe, J.M.: Lógica e Paraconsistencia, em: Novo pacto da Ciência, pp. 185–191. A Crise das Paradigmas, Anais, Escola de Comunicacão e Artes - USP (1991)
3. Abe, J.M.: On the Foundations of Annotated Logics (in Portuguese). Ph.D. Thesis, University of São Paulo (1992)
4. Abe, J.M.: On annotated modal logic. Mathematica Japonica **40**, 553–560 (1994)
5. Abe, J.M.: Some recent applications of paraconsistent systems to AI. Logique et Analyse **157**, 83–96 (1997)
6. Abe, J.M.: Curry algebra $P\tau$. Logique et Analyse **161-162-163**, 5–15 (1998)

7. Abe, J.M.: Paraconsistent Artificial Neural Networks; An introduction. In: Carbonell, J.G., Siekmann, J. (eds.) Lecture Notes in Artificial Intelligence, vol. 3214, pp. 942–948. Springer, Heidelberg (2004)
8. Abe, J.M. (ed.): Paraconsitent Inteligent Based-Systems. Springer, Heidelberg (2015)
9. Abe, J.M., Akama, S.: Annotated logics $Q\tau$ and ultraproduct. Logique et Analyse **160**, 335–343 (1997) (published in 2000)
10. Abe, J.M., Akama, S.: On some aspects of decidability of annotated systems. In: Arabnia, H.R. (ed.) Proceedings of the International Conference on Artificial Intelligence, vol. II, pp. 789–795. CREA Press (2001)
11. Abe, J.M., Akama, S.: Annotated temporal logics $\Delta\tau$. In: Advances in Artificial Intelligence: Proceedings of IBERAIA-SBIA, LNCS, vol. 1952, pp. 217–226. Springer, Berlin (2000)
12. Abe, J.M., Akama, S., Nakamatsu, K.: Monadic curry algebras $Q\tau$. In: Knowledge-Based Intelligent Information and Engineering Systems: Proceedings of KES 2007—WIRN 2007, Part II, pp. 893–900, Lecture Notes on Artificial Intelligence, vol. 4693 (2007)
13. Abe, J.M., Akama, S., Nakamatsu, K.: Introduction to Annotated Logics. Springer, Heidelberg (2016)
14. Akama, S., Abe, J.M.: Many-valued and annotated modal logics. In: Proceedings of the 28th International Symposium on Multiple-Valued Logic, pp. 114–119, Fukuoka (1998)
15. Akama, S., Abe, J.M.: Fuzzy annotated logics. In: Proceedings of IPMU'2000, pp. 504–508, Madrid, Spain (2000)
16. Akama, S., Abe, J.M., Nakamatsu, K.: Constructive discursive logic with strong negation. Logique et Analyse **215**, 395–408 (2011)
17. Ávila, B.C.: Uma Abordagem Paraconsitente Basea da em Logica Evidencial para Tratar Excecoes em Sistemas de Frames com Multipla Heranca (in Portuguese). Ph.D. Thesis, University of São Paulo (1996)
18. Avila, B.C., Abe, J.M., Prado, J.P.A: ParaLog-e: A paraconsistent evidential logic programming language. In: Proceedings of the 17th International Conference on the Chilean Computer Society, pp. 2–8. IEEE Computer Society Press, Valparaiso (1997)
19. Blair, H.A., Subrahmanian, V.S.: Paraconsistent logic programming. Theor. Comput. Sci. **68**, 135–154 (1989)
20. Carvalho, F.R., Abe, J.M.: Tomadas de Decisão com Ferramentas da Lógica Paraconsistente Anotada (in Portuguese), Editora Edgard Blucher Ltda (2011)
21. da Costa, N.C.A., Abe, J.M., Subrahmanian, V.S.: Remarks on annotated logic. Zeitschrift für mathematische Logik und Grundlagen der Mathematik **37**, 561–570 (1991)
22. da Costa, N., Prado, J., Abe, J.M., Ávila, B., Rillo, M.: Paralog: Um Prolog paraconsistente baseado em Logica Anotada, Colecao Documentos, Serie Logica e Teoria da Ciencia, IEA-USP, n° 18 (1995)
23. da Costa, N.C.A., Subrahmanian, V.S., Vago, C.: The paraconsistent logic $P\mathcal{T}$. Zeitschrift für mathematische Logik und Grundlagen der Mathematik **37**, 139–148 (1991)
24. Da Silva Filho, J.I.: Métodos de interpretação da Lógica Paraconsistente Anotada com anotação com dois valores LPA2v com construção de Algoritmo e implementação de Circuitos Eletrônicos (in Portuguese), Ph.D. Thesis, University of São Paulo (1999)
25. Da Silva Filho, J.I., Abe, J.M.: Emmy: a paraconsistent autonomous mobile robot. In: Abe, J.M., Da Silva Filho, J.I. (eds.) Frontiers in Artificial Intelligence and its Applications, pp. 53–61. IOS Press, Amsterdam (2001)
26. Lopes, H.F.S., Abe, J.M., Anghinah, R.: Application of paraconsistent artificail networs as a method of aid in the diagnosis of Alzheimer disease. J. Med. Syst. 1–9 (2009)
27. Mario, M.C., Abe, J.M., Ortega, N., Jr, Del Santo, M.,: Paraconsistent neural network as auxiliary in cephalometic diagnosis. Artif. Org. **34**, 215–221 (2010)
28. Nakamatsu, K., Abe, J.M., Suzuki, A.: Annotated semantics for deefeasible deontic reasoning, pp. 470–478. Rough Sets and Current Trends in Computing, Lecture Notes in Artificial Intelligence series (2000)
29. Nakamatsu, K., Abe, J.M., Akama, S.: Intelligent safety verification for pipeline process order control based on bf-EVALPSN. In: ICONS 2012: The Seventh International Conference on Systems, pp. 175–182 (2012)

30. Nelson, D.: Constructible falsity. J. Symb. Logic **14**, 16–26 (1949)
31. Prado, J.P.A.: Uma Arquitetura em IA Basea da em Logica Paraconsistente. Ph.D. Thesis, University of São Paulo (2006)
32. Subrahmanian, V.: On the semantics of quantitative logic programs. In: Proceedings of the 4th IEEE Symposium on Logic Programming, pp. 173–182 (1987)
33. Suppes, P.: The axiomatic method in empirical science. In: Henkin, L. (ed.) Proceedings of the Tarskian Symposium, pp. 465–479. American Mathematical Society (1974)

Printed in the United States
By Bookmasters